ELEMENTS

OF

GEOLOGY,

INTENDED FOR THE USE OF STUDENTS.

BY

SAMUEL ST. JOHN,

PROFESSOR OF CHEMISTRY AND GEOLOGY IN
WESTERN RESERVE COLLEGE.

Chimborazo.

"Geology, in the magnitude and sublimity of the objects of which it treats, undoubtedly ranks next to Astronomy in the scale of the Sciences."

Sir J. F. W. Herschel.

Fifth Edition.

NEW YORK:
G. P. PUTNAM & CO., 10 PARK PLACE.
M.DCCC.LIV.

WILLIAM H. SHAIN,
HUDSON STEREOTYPE FOUNDRY.

PREFACE.

THE following treatise has been prepared in accordance with the request of a large number of teachers, who desired to introduce the study of Geology into our higher schools. The work is designed to be strictly elementary, and hence does not embrace protracted discussions on the more abstruse and undetermined problems of theoretical geology; it aims to engage the interest of the pupil in the facts of the science.

The authorities for the facts cited are principally Lyell, Murchison, Buckland, Ansted, Agassiz, Hitchcock, Dana, and the State Geological Surveys, particularly that of Professor Hall of New York.

The engravings were executed by Professor Brainerd, and in some instances the subjects were sketched from nature by him.

An analytical table of contents seemed to the author preferable, for the purposes of examination and review, to a series of questions appended to each page.

Experience in teaching geology indicates that nothing so much encourages the pupil and facilitates his progress, as frequent use of sections on a large scale, to illustrate both the actual modes of occurrence of geological phenomena, and theoretical views. The map recently prepared by Professor Hall is admirably adapted to this purpose.

ANALYTICAL TABLE OF CONTENTS.

DESIGNED AS A BASIS FOR REVIEW AND EXAMINATION

CHAPTER II.

THE STRUCTURE AND POSITION OF ROCKS.

CHAPTER III.

PALÆONTOLOGY.

CHAPTER IV.

THE UNSTRATIFIED ROCKS.

CHAPTER V.

STRATIFIED PRIMARY ROCKS.

CHAPTER VI.

PALÆOZOIC ROCKS.

CHAPTER VII.

ROCKS OF THE SECONDARY PERIOD.

CHAPTER VIII.

ROCKS OF THE TERTIARY PERIOD.

CHAPTER IX.

ROCKS OF THE QUATERNARY PERIOD.

CHAPTER X.

THEORETICAL GEOLOGY.

CHAPTER XI.

PRACTICAL GEOLOGY.

CHAPTER XII.

HISTORY OF GEOLOGY.

CHAPTER XIII.

RELATION OF GEOLOGY TO RELIGION.

CHAPTER XIV.

GEOGRAPHICAL GEOLOGY.

GEOLOGY.

DEFINITION AND OBJECT OF GEOLOGY.

1. Geology is the Science which treats of the constitution and structure of the Earth.

Its object is to observe and describe the mineral masses, and the remains of organized bodies, animal and vegetable, which compose the globe; trace the successive changes they have undergone and discover the various laws that govern such changes.

Descriptive Geology exhibits the facts of the science;

Theoretical Geology attempts to account for them; and

Practical Geology shows their application to practical purposes.

Subservient to Geology are—*Chemistry*, which treats of the ultimate particles of matter, and their modes of combination; *Mineralogy*, which characterises and classifies the various minerals of which the earth is composed; *Botany*, and *Zoology*, which describe plants and animals; and *Physical Geography*, which relates the facts concerning the general distribution of matter at the surface of the earth—the forms and extent of continents and islands, river and mountain systems; together with the changes now occurring in them.

CHAPTER I.

GENERAL CONSTITUTION AND STRUCTURE OF THE EARTH.

Fig. 1.—The Earth, as seen from the Moon.

PLANETARY RELATIONS.

2. THE Earth is one of the planetary bodies constituting the Solar system—the third in order from the Sun—completing its circuit of 600,000,000 miles around that luminary in a year, and revolving upon its axis once a day. Its relations to the sun and other members of the Solar system determine its position in space, the amount of heat and light it receives, and consequently its vegetable and animal economy. A change in the position of its axis would alter its climate and the distribution of land and sea.

FIGURE.

3. The *form* of the Earth is that of an oblate **spheroid.** The equatorial diameter is twenty-six and a half **miles** longer than the polar.

The equatorial diameter		=7925.6 miles,
" polar	"	=7899.1 "
Difference or flattening, . .		26.5 miles.

This gives a compression of thirteen and a quarter miles to each hemisphere. If the earth should cease to rotate on its axis, the waters of the ocean about the equator would flow towards the poles, seeking the lowest level, *i. e.* the position nearest the center of the earth. The direction of rivers running towards the equator would be reversed.

This form of the globe seems to indicate that its particles have been free to obey the centrifugal force. The other planets exhibit spheroidal forms. The equatorial diameter of the planet Jupiter, exceeds the polar diameter by more than six thousand miles.

DENSITY.

4. The *density* of the Earth is five and a half times that of water. It weighs five and a half times as much as a globe of water of the same size. This is more than twice the density of the most prevalent rocks at the surface. Hence it appears that the density increases towards the center. By the pressure exerted beneath the surface the bulk of bodies is compressed, and their density consequently increased. At the depth of thirty-four miles, air would become as dense as water; at three hundred and sixty-two miles, water as heavy as quicksilver; and at the center the average minerals of the surface would be compressed into

less than one-tenth their present bulk. This would make the mean density of the globe much greater than it is. Consequently the materials within must be different from those at the surface, or this compressing force must be counteracted by some expansive power. The density of the Earth is ascertained by astronomical processes; observing the deflection of the plumb-line caused by a mountain of known dimensions, the attraction of other planets, &c.

TEMPERATURE.

5. The *temperature* of the globe is determined by a variety of influences. The temperature of the *surface* is influenced by its latitude, being warmer near the equator, and diminishing toward the poles; by its power of absorbing and retaining, or reflecting the heat of the sun's rays; by its elevation above the general level, the temperature falling as we ascend; and by the distribution of land and water, the climate of the ocean and of islands being milder and more uniform than that of continents in the same latitude. Lines joining places having the same mean annual temperature, are called *isothermal* lines. These, on the ocean, are very nearly parallel to the equator and to each other, but over continents are modified by the extent and elevation of the land. The following table gives the latitude of isothermal lines on the American and European coasts:

Isothermal Line.	Latitude on American coast.	Latitude on European coast.
77° Fahr.	24° 21′	18° 49′
68° "	32° 20′	31° 27′
59° "	38° 24′	41° 33′
50° "	41° 30′	52° 3′
41° "	44° 51′	60° 7′
32° "	51° 57′	66° 48′

Fig. 2.

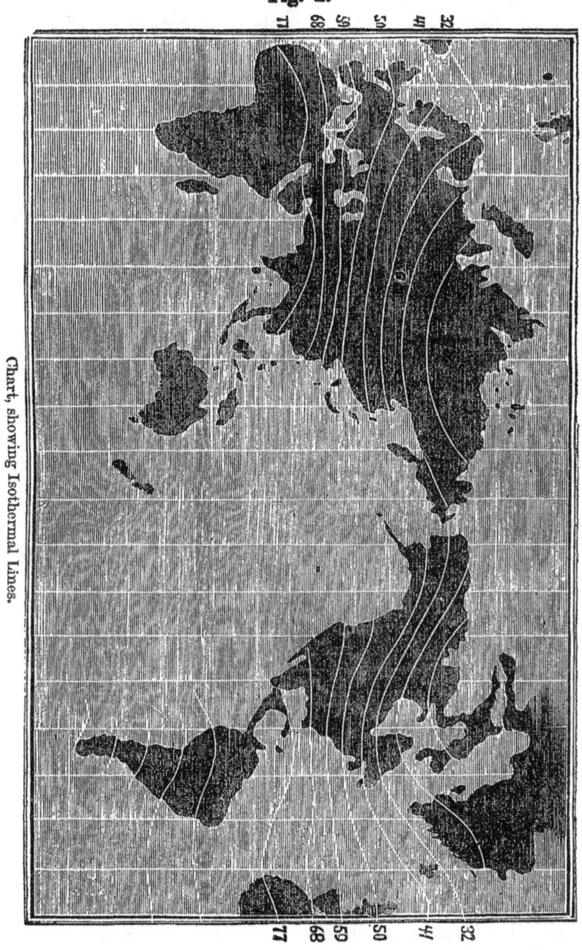

Chart, showing Isothermal Lines.

The mean temperature of the whole surface of the Earth is estimated to be 58°. The surface temperature determines the distribution and growth of plants and animals.

The heat of the sun penetrates the crust of the earth to a limited extent, rarely exceeding seventy feet. At this limit the temperature remains constant through the year. On descending below this point the temperature uniformly rises about 1° of Fahrenheit for every fifty-four feet of depth. The rate of increase varies with the nature of the rocks passed through, from 1° for thirty feet to 1° for seventy feet. This has been established by numerous experiments in mines, Artesian wells,* and mineral springs.

In the mine at New Salzwerk, near Minden, in Prussia, two thousand feet deep, the increase of temperature was at the rate of 1° for fifty-four feet. The mean result of a large number of observations in the Saxony mines, gives an increase of 1° for seventy-six feet at depths of two

* Artesian wells are borings through which the water rises nearly or quite to the surface, where no indication of springs existed. They are so called from the French province Artois—the ancient Artesium—where they were used as early as the 12th century of the Christian era.

Fig. 3.

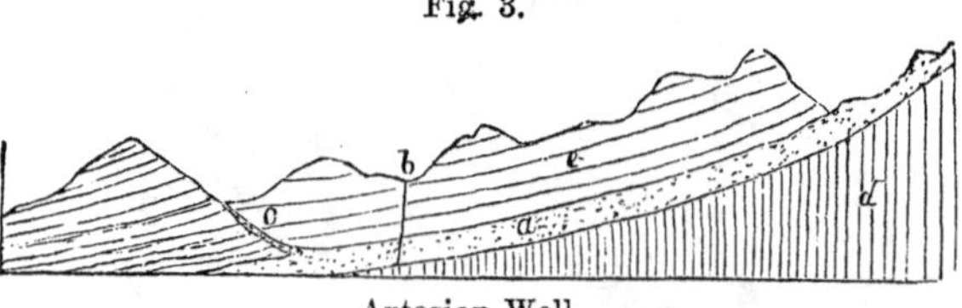

Artesian Well.

a. porous water-bearing rock between the rocks *e* and *d*, which are impervious to water. Through the well at *b*, the water rises as in a syphon.

thousand feet; while in a coal mine in Durham, at about the same depth, it is 1° for fifty-nine feet. The rate of increase observed in sinking the well at the Barriere de Grenelle, Paris, was 1° for fifty-eight feet. In a very deep Artesian well recently sunk at Mondorf, on the frontier of France and Luxembourg, to a depth of nearly two thousand three hundred feet, the water at two thousand two hundred feet had a temperature of 93° Fahrenheit, showing an increase at the rate of 1° Fahrenheit for fifty-four feet.* With this rate of increase, at the depth of fifty miles, the heat would be sufficient to melt all known rocks.

SURFACE CONFIGURATION.

6. The *surface outline* of the Earth is very irregular, intersected by mountains and vallies, seas and rivers. The highest mountains exceed five miles.

Dhawalagiri,	in the	Himalayas,	28,072 feet.
Aconcagua,	"	Andes,	23,200 "
Chimborazo,	"	"	21,420 "
Illimani,	"	"	21,149 "
Mount Blanc,	"	Alps,	15,743 "
Pic Nethou,	"	Pyrenees,	11,168 "

The mean height of all the solid parts of the Earth's surface above the ocean, is estimated by Humboldt at about one thousand feet; that of South America, one thousand one hundred and fifty one; of North America, seven hundred and forty-eight; of Europe, six hundred and seventy-one; and of Asia, one thousand one hundred and thirty-two feet. The length of the chain of mountains between Siberia and India is ten thousand miles, and its breadth one thousand five hundred miles. The mountains of Amer-

* Ansted.

ica, the Andes and Rocky Mountains, extend over 120° of latitude. There are also vast depressions below the general level. The Caspian and Aral seas are situated in such a depression, whose whole area is not less than one hundred thousand square miles. The surface of the Caspian sea is eighty-three and a half feet below the level of the ocean, and the sea has a depth of six hundred feet. The lake of Tiberias has its surface four hundred and sixty-six feet below the Mediterranean, while that of the Dead Sea is one thousand three hundred and eighty-eight feet below the same level, and its bottom in some places three hundred fathoms lower. But the great deeps of the Earth's surface are occupied by the ocean. Soundings in the Atlantic ocean (27° south latitude 17° west longitude) gave a depth of fourteen thousand five hundred and fifty feet, and four hundred and fifty miles west of Cape Good Hope, sixteen thousand and sixty two feet; while in latitude 15° south, longitude 23° west, a line of twenty-seven thousand six hundred feet, did not reach the bottom. Humboldt estimates the mean depth of the ocean at one thousand feet.

7. The *distribution of Land and Water* upon the surface of the earth is very unequal.

Water occupies	145,300,000	square miles,
Land, "	51,500,000	" "
Total, ...	196,800,000.	

Nearly three-fourths of the surface is, therefore, beneath water, and there is sufficient water to submerge the whole.

THE MEANS OF GEOLOGICAL INVESTIGATION.

8. By the *Crust of the Earth*, is understood, the exterior portion of it, which is accessible to man, a few miles in thickness. The following diagram represents the propor-

tions which the crust, with its mountains and seas, bears to the globe. The point *i* indicates a depth of one hundred miles; *k* five hundred miles. The divergence of the lines at *a* and *h* shows the depth of the ocean. The prominences at *d*, *e*, *f*, and *g*, adequately represent the height of mountains. The line *l* is at forty-five miles above the surface—the supposed limit of the atmosphere. The thickness of the heavy line representing the crust is one four hundredth part of the distance to the center—*i. e.* ten miles.

Fig. 4.

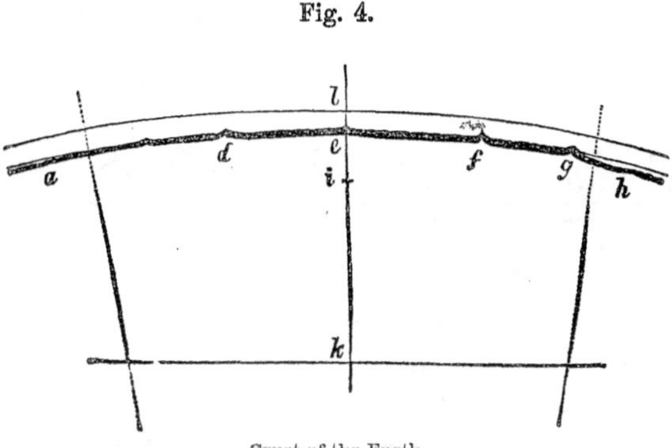

Crust of the Earth.

This crust, insignificant as it may seem, when compared with the mass of the globe, since it embraces all the stratified rocks, enables us to determine the various changes which the Earth has undergone.

9. The *means of investigating Geological phenomena* are found principally in the fracture and dislocation of the layers of the crust of the Earth. Artificial excavations and borings, aside from the inclination of these layers, are of little avail. The deepest mine extends but a fraction of a mile below the surface. Had the strata remained, as they

were deposited, in a horizontal position, we should have been enabled to investigate only the few superficial ones; but the slanting position into which they have been thrown, brings the lowest of them to the surface, and presents their edges for examination, as in figure 5, in which the layers *a*, *b*, *c*, *d*, and *e*, present themselves to the observer at the surface, successively from the superficial to the deep seated. Other masses of the Crust of the Earth have been pressed up, in a melted state, from deep recesses to the surface, and are thus subjected to inspection, as lavas of volcanos.

Fig. 5.

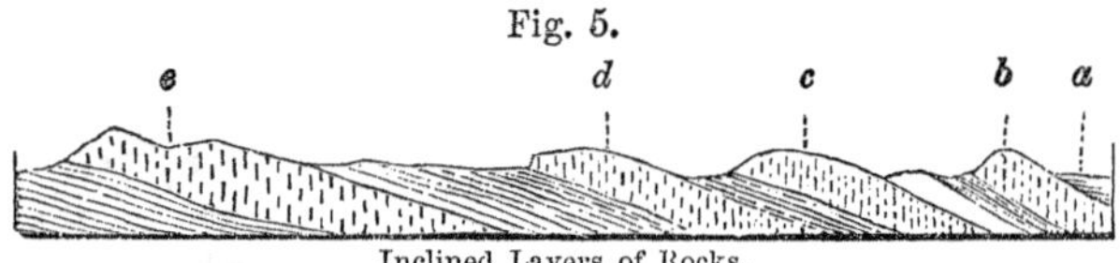

Inclined Layers of Rocks.

CHEMISTRY OF THE EARTH.

10. Chemists recognize sixty elements, or simple bodies, whose combinations produce all the varieties of matter with which we are acquainted. Many of these occur in small quantities, and are rarely seen. Fifteen or sixteen of these elements enter largely into the composition of rocks.

These substances, however, very rarely present themselves in their elementary state; but, combined with each other, they make the greatest portion of the Crust of the Earth.

The most prevalent of these is *Oxygen*, which forms eight-ninths of water, one-fifth of the atmosphere, and constitutes one-half of all the matter known to us. With silicon it forms silica; with potassium, potassa; with iron the oxide of iron, &c. Few minerals occur which do not contain oxygen.

Hydrogen forms a portion of minerals, especially bituminous coal, and enters into the composition of water.

Nitrogen is not so abundant, but is found in the bones of animals, living and fossil, in vegetables and in the atmosphere.

Carbon is the most abundant ingredient in coal, and enters into the composition of limestone, which is the carbonate of lime.

Sulphur exists in the sulphurets of the metals: sulphuret of iron—iron pyrites; sulphuret of lead—galena, or lead ore; also in the sulphates, as sulphate of lime—gypsum, or plaster of Paris. It is thrown out extensively by volcanos.

Chlorine is one of the constituents of rock salt (chloride of sodium) and is widely diffused in the ocean.

Fluorine occurs in fluoride of calcium, (fluor spar,) and other minerals.

Phosphorus enters into the composition of many minerals, and of animal bones, as the phosphate of lime.

Silicon exists in most of the rocks, combined with oxygen, as silica—quartz—which constitutes forty-five per cent. of the Crust of the Earth.

The oxide of *aluminium*—alumina forms one-fifth of the mineral feldspar, and abounds in clay and slate rocks. It is estimated at ten per cent of all the rocks.

The oxide of *Potassium* also enters largely into the composition of feldspar and clay.

Sodium forms a part of rock salt and other minerals.

The oxide of *calcium* (lime) occurs chiefly as a carbonate (limestone, marble,) which is estimated to form fourteen per cent. of the globe's crust.

Magnesia, the oxide of magnesium, enters into the

composition of many rocks, and abounds in the magnesian limestone.

Iron is very widely diffused, in the various forms of its ores, oxide, carburet, sulphuret, &c.

The combinations of *manganese*, also are common ingredients of rocks in small quantities.

Some knowledge of chemical facts facilitates the researches of the geologist, enabling him to perceive many changes which have taken place in the rocks through the agency of chemical affinity.

MINERALOGY OF GEOLOGY.

11. Of the hundreds of simple minerals described by mineralogists, a very few occur in large quantities, making up the great mass of the rocks. The most prevalent are the following:

Quartz, a hard mineral, striking fire with steel; and scratching glass and other substances, (a few rare gems excepted;) infusible before the blow-pipe, and insoluble in ordinary acids. The transparent grains of granite, and of sandstone, are quartz.

Feldspar, a reddish white, or flesh-colored mineral; opake, easily scratched, becoming white and glassy by the heat of the blow-pipe.

Mica, a glistening mineral, tough and elastic, susceptible of division into minute laminæ, giving to rocks in which it abounds a silvery aspect.

Hornblende, a black, or dark green mineral, of glassy luster, occurring in granite instead of mica, or associated with it, when the rock is called syenite; in trap rocks and in hornblende-slate.

Augite is nearly identical with hornblende in all its characteristics.

Talc, a very soft mineral, of pearly luster, and greasy feel; the characteristic ingredient of steatite or soapstone, French chalk, and chlorite.

Serpentine, a green, opake mineral, becoming yellow ish gray on exposure; it sometimes presents a mottled appearance.

Carbonate of Lime, marble, chalk, limestone; effervescing with acids, and reduced by heat to caustic lime.

Dolomite, magnesian limestone; a compound of the carbonate of lime with the carbonate of magnesia, infusible, and effervescing slowly with acids; used in the manufacture of Epsom salts.

Sulphate of Lime, gypsum, or plaster of Paris, alabaster.

Chloride of Sodium, common salt.

Tourmaline—sometimes called *schorl*—a black, or dark brown mineral; brittle, becoming electrical when heated.

Iron Pyrites, a yellow, hard, cubical mineral, often mistaken for gold; composed of sulphur and iron.

Iron Ores, and *Coal*.

While an intimate knowledge of Mineralogy is not indispensable to the Geological student, an acquaintance with those minerals which enter largely into the composition of rocks is necessary, since the structure of the rocks is determined by them.

AGENCIES USED IN EFFECTING GEOLOGICAL CHANGES.

12. The study of Nature in all its departments, impresses upon the mind the idea of incessant change, as well in the solid strata of the earth, as in the more fleeting forms of animal and vegetable organization. Matter and motion seem to be inseparably connected. Changes, however, whether slight and momentary, or so grand as to task the imagination in their conception, and require ages for their

completion, evince the most enduring permanence in the laws which produce and govern them; and while our belief in the constancy of natural laws is intuitive, we perceive that incessant change is the means of effecting stability in the whole system.

13. While Geologists agree respecting the *nature* of the agents employed in effecting former geological changes —that they were identical with those now in operation— they are divided in opinion with reference to the *intensity* with which these agents operated. One class contends that the phenomena require, for their explanation, much more extensive and violent action than we witness at present; while others claim that existing causes have operated through all periods with the same degree of energy, and when long continued are adequate to the production of all the phenomena. The latter opinion is maintained by Sir Charles Lyell in his Principles of Geology.

The agencies of geological changes may be classified as *atmospheric, aqueous, igneous,* and *organic.*

ATMOSPHERIC AGENCIES.

14. Atmospheric agents produce both *chemical* and *mechanical* effects. The atmosphere consists of nitrogen seventy-nine parts, and of oxygen twenty-one parts in a hundred; of carbonic acid about one part in a thousand; and of watery vapor a variable quantity.

Many minerals are acted upon chemically by oxygen and carbonic acid. The sulphuret of iron (iron pyrites) when exposed to moist air, undergoes decomposition; the sulphur unites with a portion of oxygen, forming sulphuric acid, and the iron with another portion of the oxygen, forming oxide of iron; the two new bodies, thus formed, unite with each other, and the sulphuret of iron is converted into the sulphate of iron, (copperas.) Carbonic acid unites with the oxide of iron, producing the carbonate of iron. It also renders water a solvent of limestone, so that water charged with it, falling upon a limestone rock dissolves, and thus removes portions of it, leaving fissures and caverns in the rock. These chemical changes are usually followed by a crumbling—*disintegration*—of the rocks. It is, however, sometimes the means of consolidation, as when the carbonate of lime in solution infiltrates a mass of loose sand, it cements it, producing a calciferous sandstone.

15. *Winds* modify the surface of the earth by drifting sand, gravel, shells, &c., from exposed to sheltered positions, in deserts and on the sea coasts.

The sands of the Lybian desert have buried ancient cities and temples in Egypt. Many parts of the coasts of England, Holland and France are partially or wholly submerged by them. In Brittany, of a whole village overwhelmed by drifting sands, nothing is visible above their surface but the spire of a church. In Cornwall and Suffolk, England, are sand hills composed partly of comminuted shells, several hundred feet above the level of the sea, advancing upon the cultivated land at the rate of five miles in a century. Ansted states that on the coast of Spain, at Cape Finisterre, they have advanced sixteen miles in half a century—five hundred and sixty yards per annum. These sand hills are called in England *dunes* or *downs;* in Scotland, *links.* They are sometimes arrested in their progress by the roots of plants, especially of the *arundo arenaria,* which bind them into a firm mass. They are also sometimes indurated by the carbonate of iron, or lime, or by silica, cementing them, thus forming sandstones. The phenomena of drifting sands are exhibited in our country on the sandy capes—as at Cape Cod.

16. *Frost* exerts a destructive influence upon rocks. When water which has entered the fissures or pores of rocks freezes, it expands with a force adequate to rend the rocks; each fragment is again subjected to the same process until it is reduced to powder. This is most conspicuous in rocks of loose texture, but the densest and firmest do not escape. Blocks and ledges of granite, by this process of exfoliation, lose their angles and edges, and are oftentimes converted into rounded and grotesque forms. The accumulation of fragments at the bases of precipices is called *a talus*, usually presenting a slope of about 40°.

AQUEOUS AGENCIES.

17. *Aqueous agencies* are exerted either *chemically*, dissolving and decomposing rocks, or *mechanically*, abrading them and removing their particles. Water is the most general solvent known, and few rocks at the surface escape its influence. Its solvent power is enhanced by acids and by heat.

The amount of common salt, carbonate of lime, and other salts, held in solution in the waters of the Earth, is enormous. The amount of solid matter dissolved in the ocean is estimated to be more than thirty-nine parts in a thousand of water. From this source, shell fishes, coral-polyps and other animals inhabiting the sea, obtain their supplies for their skeletons—shells, coral-reefs, &c. The amount thus withdrawn is replaced by the solvent power of the water exerted upon the rocks to which the ocean has access. Water charged with carbonic acid, having dissolved the carbonate of lime, or the oxide of iron, may lose its solvent power over these bodies by the removal of the acid; in such case the contained salt is precipitated. Such deposites of carbonate of lime are called *tufa* or *travertin;* or, if they occur in water trickling from the crevices of caverns, they are called *stalactites.* The deposits made on the floor beneath the stalactites, are called *stalagmites.* Similar deposites of the salts of iron previously held in solution constitute the *hydrate of iron, bog iron ore.* The waters of hot springs often hold *silica* in solution, which they deposite on cooling.

18. The *mechanical* agency of water is still more manifest, and is either gradual, as in the wearing down rocks by the rain, or rapid, by torrents, and ocean storms.

Rivers are most efficient agents in transporting mineral masses. When their fall is considerable and their motion rapid, they wear their beds, deepening their channels. They deposit the materials suspended in them along their banks, upon their bottoms, or at their mouths

Their eroding power is increased by the friction of the ice, sand, and pebbles conveyed by them, and is exhibited on a grand scale at large waterfalls. The falls of Niagara are one hundred and fifty feet in height, and the average amount of water passing over each minute, is estimated at six hundred and seventy thousand tons. The position of the falls is not stationary; they have receded about fifty yards in forty years, and it is difficult to avoid the conclusion, that they were originally at Queenstown, seven miles below their present position. The length of time required to wear through this space, can not be satisfactorily determined, since they must have passed through several rocks varying in texture, in which the rate of wearing can not now be ascertained. For the same reason their future progress can not be predicted Their present situation is favorable to rapid recession. The uppermost rock, over which the water falls, is a hard limestone ninety feet thick; beneath is a soft shale, which is easily worn away by frost and the friction of the water, leaving the limestone jutting over, in table rocks, sometimes forty feet beyond the shale. From time to time these table rocks fall off.

Fig. 6.

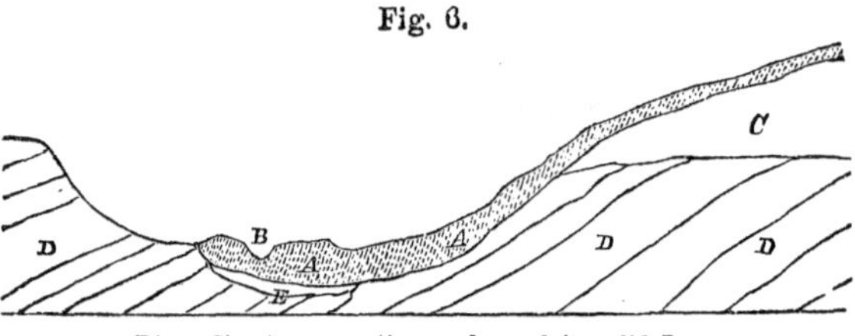

River Simeto excavating a channel in solid Lava.

19. The Simeto, one of the principal rivers of Sicily, has excavated its bed from fifty to several hundred feet wide, and forty to fifty feet deep, since A. D. 1603. In the cut given above, A, indicates the lava bed, which descending from the summit of the great volcano, has flowed five or six miles, and usurped the old bed of the Simeto, E; B, the present bed of the river. The lava is not soft and scoriaceous, but is a compact, hard rock. C, indicates the foot of the cone of Etna, and D, marine and volcanic beds. The general declivity of the river is slight, having two falls of six feet each. The abraded materials of the volcanic rock have greatly assisted the attrition.*

20. The mechanical force of rivers is exerted in carrying forward the sedimentary matter contained in there. When the rapidity of the stream is diminished the heavier portions of the sediment are deposited: but when the rate

* Lyell's Principles.

of the current is increased by freshets, coarse gravel, pebbles, and even large blocks of stone, are borne forward by it. Torrents, produced by the bursting of dams, are very destructive in their effects. A remarkable example of this action occurred in the Valais (Switzerland) in A. D. 1818. A dam four hundred feet high and six hundred feet wide was formed across the valley during the winter, of the rocks and ice brought down from the mountains by avalanches. The water of the river Dranse accumulated in this lake containing more than eight hundred million cubic feet. The inhabitants, anticipating destructive results from its eruption, opened an orifice in the dam near the surface of the water, by which three-fifths of the water was conducted off in three days, without violence. The barrier then gave way, and all the remaining waters were discharged in half an hour. The torrent reached the Lake of Geneva, forty-five miles distant, in six hours, destroying houses and trees that were in its path, and strewing the valley with rocks and earth.

The rapidity with which water excavates a channel was illustrated a few years since in the town of Glover, Vermont. A small channel made from Long Pond, which was two and a quarter miles long by one and a quarter wide, with a depth of one hundred and fifty feet, was in a few minutes so enlarged as to discharge all the water into Lake Memphremagog, twenty miles distant, with the usual violent effects of an inundation.

21. The transporting power of water is rendered more efficient by the loss of weight which minerals sustain in water. Most minerals weigh about one-half as much in water as in the air. It is ascertained by experiment that a velocity of six inches per second will raise and transport

on a horizontal surface, fine sand; eight inches, coarse sand; twelve inches, gravel; and twenty-four inches, rounded pebbles. Fresh water moving with a velocity of one and a half mile per day, raises fine clay, and eight and a half miles per day removes sand. Fine mud settles so slowly in moving water that it is often transported hundreds of miles before reaching the bottom. For this reason the sediment of rivers near their mouths consists mostly of mud.

22. Deposits of mud are called *silt.* The thickness of the alluvial bed of the Mississippi river one hundred miles above its mouth, is two hundred and fifty feet. Rivers in this way sometimes raise their beds higher than the adjoining plains, and forsaking their old beds, flow in new channels. The Po has repeatedly deserted its bed, destroying a number of towns. Several churches have been taken down and removed to escape its erratic movements. In one instance, during the fifteenth century, it returned to the bed it had forsaken, but again deserted it during the same century. It is now restrained by artificial banks on an elevated mound, so that its surface at Ferrara is higher than the roofs of the houses in the city. The Mississippi is also confined by *levees.*

23. *Deltas* are deposits of mud, sand, and gravel, made at the mouths of rivers, triangular in form, with the apex of the triangle up the stream. They are called deltas from their resemblance to the Greek letter Δ. Deltas are *fluviatile* when formed at the entrance of one river into another; *lacustrine* when at the entrance of a river into a lake; and *maritime* when the river empties into the sea.

These deltas often accumulate rapidly, and mountain lakes are sometimes entirely filled up by them. The *Rhone* and *Arve* enter the Lake of Geneva as turbid streams, but the Rhone emerges from the lake perfectly transparent. The delta formed is six miles long—a flat alluvial plain of sand and mud. An ancient town, Port Vallais, (Portus Valesiæ of the Romans,) which anciently stood upon the shore of the lake, has had its harbor silted up, and is now more than a mile and a half inland. Mr. Lyell says, "We may look forward to the period when this lake will be filled up, and then the distribution of the transported matter will be suddenly altered, for the mud and sand brought down from the Alps will thenceforth, instead of being deposited near Geneva, be carried nearly two hundred miles southwards, where the Rhone enters the Mediterranean." The Rhone after leaving the Lake of Geneva again acquires sediment, and has formed a maritime delta at its mouth of three hundred and twenty thousand acres. The mouth of the river has advanced more than eight miles since the commencement of the Christian era.

Fig. 7.

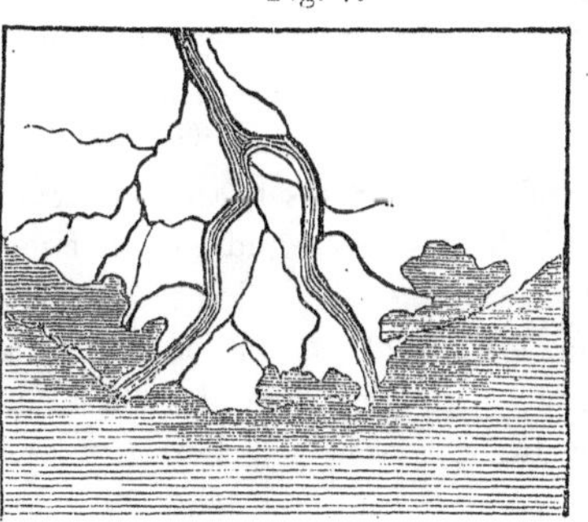

Delta of the Nile.

24. The delta of the *Nile* commences one hundred miles from the Mediterranean, and is two hundred and thirty miles broad. It does not at present increase much, because the sediment which reaches the Mediterranean is swept away by an easterly current, and the detritus brought down by its periodic floods are spread over the plains of Egypt. This fertilizing deposit, consisting of silica, alumina, oxide of iron, and carbonates of lime and magnesia, has accumulated

to the depth of more than six feet since the commencement of the Christian era.

25. The *Ganges* presents a delta of gigantic dimensions, commencing two hundred and twenty miles from the sea, and having a base line of two hundred miles. It renders the waters of the ocean turbid sixty miles from the coast. The total amount of sediment annually discharged being one four hundred and fiftieth part of the whole weight of the water, is six billion three hundred and sixty-eight million tons, an amount sufficient to cover a township five miles square to the depth of ten feet.

The whole coast of Northern Central Europe, from Calais to the Baltic, has been covered with mud brought down by the *Rhine*, which in some places is several hundred feet thick. Experiments show that the quantity yielded by this river, is about four hundred tons weight per hour.

26. The delta of the *Mississippi* comprises an area of thirteen thousand six hundred square miles, projecting into the Gulf of Mexico, with a depth of 1056 feet, and containing 2720 cubic miles. The quantity of water annually discharged by the river is 14,883,360,636,880 cubic feet; quantity of sediment discharged is 28,188,083,892 cubic feet, being the one five hundred and twenty-eighth part of the water. One cubic mile is formed in five years and eighty-one days. The formation of the whole delta would have required, at this rate, 14,203 years. The quantity annually discharged is sufficient to cover a township five miles square with a layer of mud forty feet thick.

27. The accessions to the land from the sediment brought down by the *Po* and the *Adige*, are one hundred miles long, and vary in width from two miles to twenty miles. The town Adria, a seaport in the time of the Em-

peror Augustus, which gave its name (Adriatic) to the gulf, is now more than twenty miles inland.

28. The detritus of the *Amazon* does not form a delta at its mouth, but is swept by ocean currents, and distributed partly in the ocean, where it is discernible three hundred miles from the mouth of the river, and partly to the coast of Guiana, where it has formed an immense alluvial deposit. When the matter deposited at the mouth of a river has not accumulated sufficiently to rise to the surface of the water, it is called a *bar*. Such a bar, called "the Overslough," is formed a few miles below Albany, in the still water caused by the meeting of the waters of the Hudson with the tide from the ocean. The attempt to remove this obstacle to navigation by contracting the river, tended only to increase the velocity of the current and thrust the bar farther down the river.

29. In addition to the materials conveyed mechanically by water, extensive deposits are made by *springs*. The waters of many springs are charged with carbonate of lime, which is deposited when the water issues from the rocks. Substances deposited in such water become *incrusted* with limestone. This is not petrifaction, since the substance undergoes no change, but is simply enveloped in the incrusting mineral.

Fig. 8.

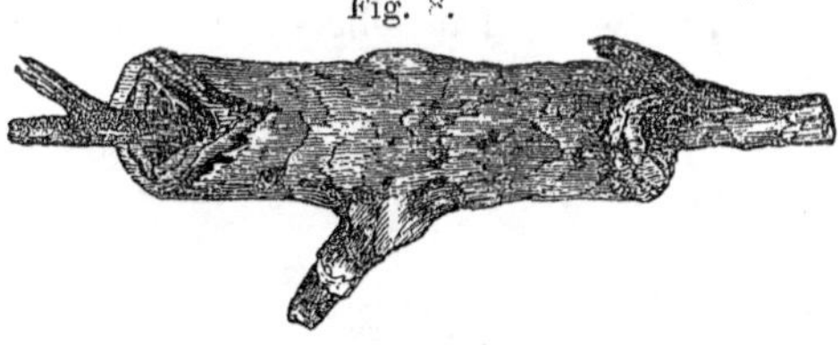

Incrusted Twig.

The preceding cut exhibits a stick thus incrusted. Many

such springs in northern Ohio incrust the mosses growing in their vicinity, furnishing beautiful but frail specimens. The loose, porous rock formed by these deposits is called *tufa;* the compact is called *travertin.* Beds of these, some miles in length and several hundred feet thick, are formed by springs in France and Italy. At San Filippo, in Italy, medallions are made by conducting the water into moulds; the deposited matter filling the mould, presents a marble cast of the figure. The beautiful alabaster of Tabreez, in Persia, has the same origin.

30. Deposits of *silicious* matter are made by hot springs. In the Azores islands hot springs, rising through volcanic rocks, deposit large quantities of *silicious sinter*, as it is called, incrusting with beautiful crystalline scales all substances with which their waters come in contact. But the most remarkable hot springs are the Geysers of Iceland, whose circular basins are lined with the *silica* of their waters. Some springs deposit iron ore, common salt, asphaltum or mineral pitch, (Seneca oil,) &c.

31. Another mode of action of water is exhibited in *land slides.* In the year 1806, after a rainy season, the Rossberg, a mountain in Switzerland, was undermined, and a mass of two thousand millions of cubic feet precipitated into the valley, forming hills two hundred feet high, and destroying several villages. Land slides have also occurred from a similar cause in the White Mountains, and near Troy, New York. The banks of the Lake and the Cuyahoga valley in the city of

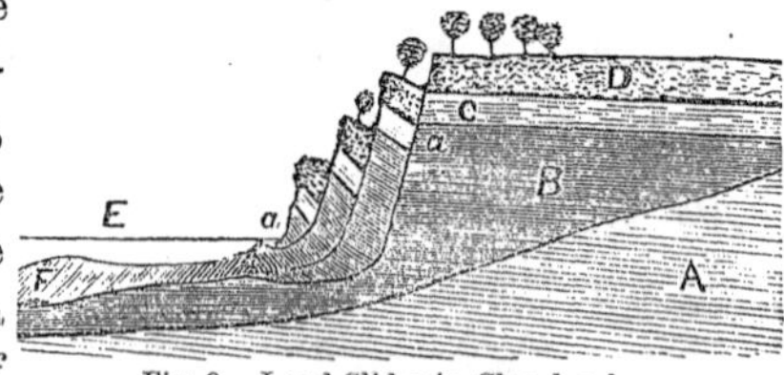

Fig. 9.—Land Slides in Cleveland.

Cleveland, are frequently subjected to the same process of degradation, as shown in figure 9. The water which falls upon the surface passes through the strata of gravel and sand D and C, and accumulates upon the bed of blue clay B, which is mostly impervious to water; from this it issues, forming springs, and carrying out the sand undermines the upper beds. This degrading influence is increased by the erosive action of the waves of Lake Erie, E, on the clay bed, as shown at B, in figure 10.

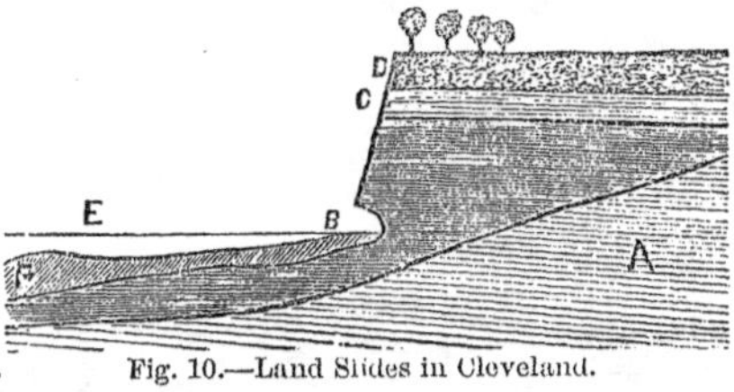

Fig. 10.—Land Slides in Cleveland.

32. *Glaciers* and *icebergs*—water in a solid form—are very influential in effecting changes upon the surface of the earth, and their peculiar phenomena have recently attracted much attention on the part of geologists.

Glaciers are immense bodies of ice formed in the valleys or on the sides of mountains, extending many miles, and remaining undissolved by the heat of summer. Por-

Fig. 11.

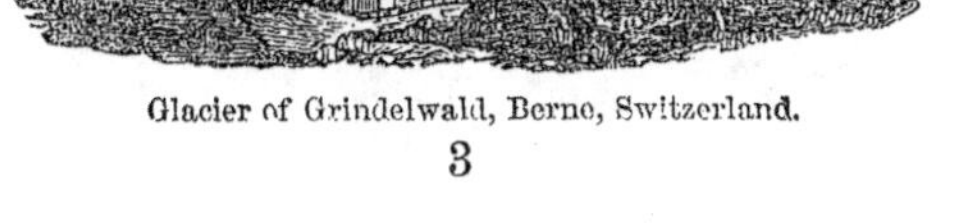

Glacier of Grindelwald, Berne, Switzerland.

tions of their mass are granular snow cemented by ice. They vary in thickness from one hundred to eight hundred feet, and have a very gentle slope. Their surface is very rough and often studded with conical masses of ice, called *needles*. They are traversed by wide fissures pro-

Fig. 12.

Glacier of Viesch, with medial and lateral moraines.

duced by contraction in the winter. Each winter adds a new layer to their surface; the snow, however, disappears from their surfaces in summer, as regularly as from the surrounding rocks. Expansion, gravity, and the forms of the valleys, cause them to advance slowly down, sometimes even to the borders of cultivation. "The very huts of the peasantry," says Professor Forbes, "are sometimes invaded by this moving ice, and many persons now living have seen the full ears of corn touching the glaciers, or gathered ripe cherries from the tree with one foot standing on the ice." The surfaces and mass of glaciers abound with fragments of rocks which are often arranged in long lines, and are called *moraines*. The rocks over which glaciers pass are smoothed, scratched and grooved, by sand and angular

Fig. 13.

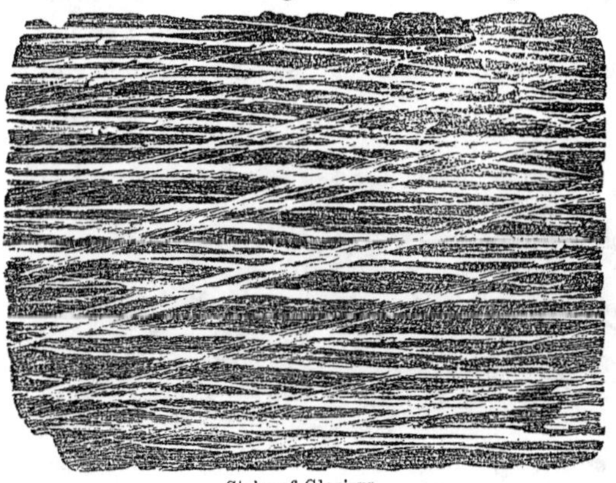

Striæ of Glaciers.

stones forced forward by their enormous pressure. These striæ and grooves are parallel to each other, because the fragments which produced them are fixed in the bottom of the glacier. Sometimes loose stones give rise to irregularities of direction and figure.

33. From glaciers formed in high latitudes near the sea, fragments of various sizes fall into the ocean, and constitute *icebergs*. The polar seas abound with them at all seasons, and marine currents float many of them into lower latitudes. They are sometimes of great size; one measured thirteen miles in length, and one hundred feet above the water, giving a thickness of seven hundred or eight hundred feet.* Floating into warmer water and atmosphere, icebergs melt, depositing their loads of earth and rocks on the bottom of the ocean. Many of them strand upon the Newfoundland Banks, and after heavy rolling and disturbance of the bottom, they melt, and mingle the fragments they conveyed with the sands of the Banks. Similar phenomena, on a smaller scale, are witnessed in lakes and rivers; a fragment of rock is floated to one side and there left by

Fig. 14.

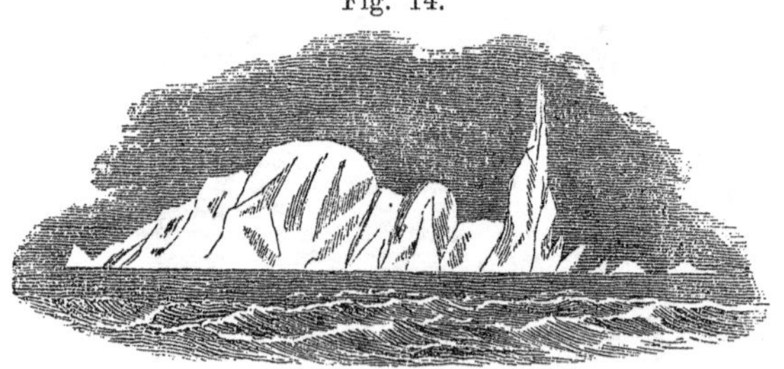

ceberg seen off Cape Good Hope, April A. D. 1829, two miles in circumference and 150 feet high.

* The specific gravity of ice is such as to make it float with one-ninth of its bulk above the surface of the water; but icebergs are porous, and float with the largest extremity immersed, so that one-seventh or one-eighth is above the water.

the melting ice; the succeeding season it may be returned by the same agency to the other side.

34. The geological agency of the *ocean* is, like that of rivers, twofold. 1st, Erosive, wearing away the coast, undermining and excavating cliffs; and, 2nd, transporting and accumulating the detritus to form new land. It produces these effects by means of *waves*, *tides*, and *currents.*

35. The action of *waves* is incessant, but varies in extent according to the nature of the exposed shore. Since they do not penetrate very deep, and have no progressive motion in the open sea, they do not affect the bottom of the ocean, except where it is very shoal. Cliffs of soft rocks, clay, chalk, or sandstone, are very rapidly undermined and worn away. The hardest rocks, however, can not resist its never ceasing attacks. Headlands of alternately hard and soft rocks, especially if intersected by crevices, are worn away with very great rapidity. The coasts of England exhibit this erosive action of the ocean in a remarkable degree. The cliffs of Yorkshire, it is ascertained by careful measurements, repeated after intervals of many years, are worn away six feet in breadth, annually. The average annual loss on the coast of Norfolk is about three feet. When Mr. Lyell, in A. D. 1829, visited Sherringham, Norfolkshire, he found water twenty feet deep, where forty-eight years before stood a cliff fifty feet high. In the county of Kent, near the mouth of the Thames, stands the Church of Reculver, upon a cliff twenty feet above the sea. In the time of King Henry VIII. the distance between the church and the brink of the cliff was one mile. The following cut represents the appearance of the spot in A. D. 1781, when the encroachments of the sea had attracted notice, though considerable space, with

other buildings, intervened between the churchyard and the cliff. The walls of an ancient Roman fortification

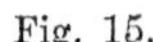

Fig. 15.

Church of Reculver in A. D. 1781.

which was two hundred and forty feet nearer the sea than the church was, had recently been undermined and precipitated into the ocean. In A. D. 1804, a part of the churchyard with some adjoining houses was washed away, and the ancient church with its two lofty spires, a well known land-mark, was dismantled, and abandoned as a place of worship. Figure 16 represents it as it appeared in A. D. 1834. It would probably long since have fallen, had not the incursions of the ocean been checked by an artificial causeway of stones and large wooden spiles driven into the sands to break the force of the waves. The isle of Sheppey, seen at some distance from the main land, on the right hand, in figure 15, which is six miles long and four broad, is continually undergoing abrasion, having lost

Fig. 16.

Church of Reculver in A. D. 1834.

fifty acres within the last twenty years.* There are other instances in which islands in this (German) ocean, and important seaport towns, have been entirely obliterated.

36. The English channel probably owes its origin to the erosive action of the ocean, the geological features of the coasts of England and France, clearly indicating that they were formerly united. During the thirteenth century, a channel half as wide as the English channel, was excavated in the north of Holland, separating Friesland from the main land. Immense labor is continually expended in Holland to prevent incursions of the ocean, which

* Lyell's Principles.

threaten to inundate much of the country, and destroy the cities of Amsterdam and Ley 'en. The effect of this constant abrasion on exposed coasts, is seen in the production of caverns, bridges, and isolated pinnacles of rock, as exemplified by the chalk "Needles" of the English coast, and the "Drongs" of the north of Scotland. Figure 17, presents a view of the cluster of rocks seen to the south of the Hillswick Ness, one of the Hebrides. These granite rocks are all that remain of a former island, which may, at an earlier period, have been a promontory of the main land.

Fig. 17.

Hillswick Ness, Hebrides Islands, Scotland.

37. The coasts of New England and Nova Scotia also exhibit striking instances of the abrading power of the ocean. Boston harbor has been formed by this agency; the outermost islands in it, exposed to the violence of the waves, consist of bare rock; and the more sheltered ones are continually losing a portion of their covering. At Cape May, on the north side of Delaware Bay, the sea has

encroached upon the land at the rate of nine feet in a year; and at Sullivan's island, near Charleston, South Carolina, four hundred and forty feet in a year. The waves of inland seas and lakes produce similar effects. The indentations of the shores of Lake Erie were ed by this agency. The mode in which it operates at Cleveland, has been already illustrated in § 33. The ordinary effects of the ocean's agency in wearing and transporting rocks, are greatly enhanced by storms. Masses of rock of from ten to thirty tons weight, have been forced by them up an inclined shore. During the erection of the Bell Rock lighthouse, six granite blocks were thrown over a rising ledge twelve or fifteen paces, and an anchor, weighing two thousand two hundred pounds, was thrown up upon the rock from a depth of at least sixteen feet.

38. The periodical elevations of the ocean waters, called *tides*, varying from two and a half feet to seventy feet in height, extend the limits of the ocean's power, and produce very marked effects in narrow channels, bays and estuaries. The action of the *tide wave* is alternate, advancing and receding, destroying as it advances, and bearing away the debris as it recedes; differing in this respect from currents, which carry the fragments which they produce or meet with, only in the direction of their course.

"The bore" is a term applied to a sudden influx of the tide into a river or strait, which, resisted by the descending water, and forced into a narrow channel, rises suddenly, and exhibits the phenomena of breakers on a shelving shore. It is most conspicuous at the time of spring or ighest tide. The bore in the Severn is sometimes nine feet high; in the Bay of Fundy seventy feet, producing inundations, sometimes sweeping off trees and animals.

39. Another kind of movement in the waters of the ocean is exhibited by *marine currents,* which perform a most important part in the economy of nature. They are of two kinds, *drift,* and *stream* currents. Drift currents are the result of a constant or prevailing wind on the surface of the ocean, penetrating to no great depth, rarely exceeding in velocity a half mile per hour, and are easily turned from their course. They are produced chiefly in the regions of the trade winds, and sometimes originate stream currents, which are immense oceanic rivers, covering a space of one to three hundred miles in breadth, and reaching to a very great depth. They are caused by the tendency of ter that is displaced to restore the equilibrium of the surface of the ocean. The origin of the displacement is not satisfactorily ascertained. It has been ascribed to winds, unequal evaporation, difference of rapidity of diurnal revolution in different latitudes, &c.

The velocity of the currents varies in different parts of their course—in some the average is sixty miles, with a maximum of one hundred and twenty miles in a day. Their temperature is either higher or lower than that of the surrounding sea, according to the temperature of the region in which they have their origin. This difference amounts to from 10° to 30° Fahrenheit. Some of these currents extend their course through many thousand miles. The same current has different names applied to it in successive parts of its course.

40. The chart, Figure 18, presents a general view of ocean currents, especially of our Gulf stream, which is a continuation of the *Mozambique* current running between the eastern coast of Africa and Madagascar: doubling the Cape of Good Hope, it enters the Atlantic ocean as the

Fig. 18.

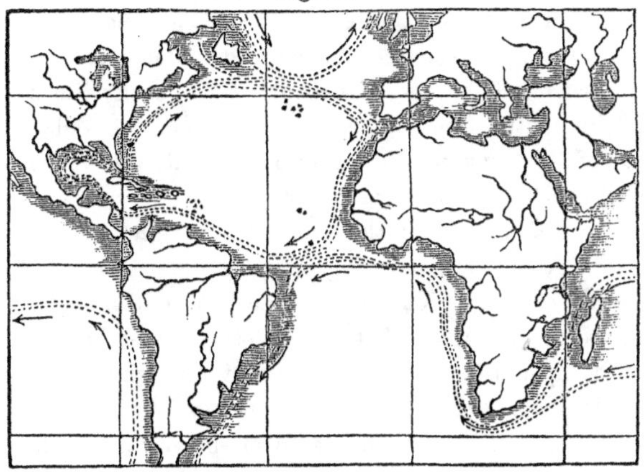

Chart of Oceanic Currents.

Cape or *Agulhas current.* Thence, under the name of *southern Atlantic current,* it flows north-easterly until its course is turned to the westward by the coast of Africa, and the opposition of the Guinea current from the north. It now forms the *Equatorial current,* and stretches across the Atlantic, on both sides of the Equator; about midway between Africa and South America it divides, sending a branch southward along the eastern coast of South America, forming the Brazil current, while the main stream continues its course by the coast of Guiana, crossing the waters of the Amazon and receiving those of the Oronoco, and enters the Caribbean Sea, as the *Guiana current.* From the Gulf of Mexico, it issues as the *Gulf stream,* running north-easterly by the coast of Florida and Cape Hatteras to the St. George and Nantucket Banks, and thence eastward by the Azores to the coast of Europe. The length of the Gulf stream from Florida to the Azores, is three thousand five hundred miles, which is traversed in seventy-eight days—at an average rate of thirty-eight miles per day. The amount of water conveyed in it is more than three thousand times the amount discharged by the Mississippi river, many times greater than all the fresh water in the rivers of the globe. In a part of its course near the Florida Gulf, its velocity is that of a torrent—five miles per hour.*

41. The temperature of the Gulf stream at the Florida coast is 86° Fahrenheit, declining, as it advances northward, to 73° at the Azores. This vast expanse of water, whose temperature is 10° above that of the ocean, must have a great effect upon the climate of adjacent countries. "A simple calculation," says Lieutenant Maury, "will show that the quantity of heat discharged over the Atlantic from the waters of the Gulf stream in a winter day, would be sufficient to raise the whole column of atmosphere that rests upon France and the British islands, from the freezing point to summer heat. It is the influence of this stream upon climate, that makes Ireland the Emerald isle of the sea, and clothes the shores of England with evergreen robes; while in the same latitude on the other side, the shores of Labrador are fast bound in fetters of ice. In A. D. 1831, the harbor of St. Johns, Newfoundland, was closed with ice in the month of June, although it is 2° farther south than Liverpool; and the influence of the Gulf stream is felt in Norway, and on the shores of Spitzbergen."

42. On issuing from the straits of Florida, the waters of the Gulf stream are of a deep indigo blue *color*, and the line of separation between it and the green waters of the Atlantic, is plainly visible for hundreds of miles. The

* Johnston's Physical Atlas.

great eddy in the middle of the Atlantic, caused by these currents, extending from 30° west longitude to the Bahamas, and between the parallels of 20° and 45° north latitude—embracing two hundred and sixty thousand square miles—is called the Sargazo sea, because its surface is covered with the *gulf weed*, (in Spanish, Sargazo.) In many places it is so thickly matted, as to retard the progress of vessels through it.

43. The *Arctic current*, originating within the polar circle, runs by the shores of Greenland and Labrador forming the *Hudson Bay current*; comes in collision with the gulf stream at Newfoundland, where it divides, and sends a branch southward by the coast of the United States, constituting the *counter current*, between the gulf stream and the coast. It enters the Caribbean sea as an under current, replacing the warm water sent through the gulf stream, and mitigating the climate of Mexico and Central America. The temperature of the Caribbean sea at the depth of 240 fathoms, has been found as low as 48° while that of the surface was 85°. This current conveys icebergs from the polar seas. Captain Scoresby counted five hundred icebergs in it at one time. Meeting the gulf stream at Newfoundland, they deposit enormous loads of rocks and earth, having thus greatly extended, and probably originated the Banks. The *depth* of the great oceanic stream currents has not been generally ascertained, but is in some cases seventy fathoms and probably more. Their mechanical effects, especially when they move rapidly, must be great at considerable depths, and they transport materials hundreds of miles.

44. As tides and currents powerfully co-operate with waves in destroying rocks on the shore of the ocean, so also do they conspire in *reproducing* land, forming banks, and

silting up estuaries. Currents assort the materials transported by them, depositing the heavy fragments of rock very near where they receive them, conveying the sand to a greater distance, while the mud settling very slowly may be transported a very great distance. Since the direction and velocity of the current are tolerably uniform, the deposits will be in a good degree homogeneous. The bed of the German ocean is traversed by sand banks, one of which "the Dogger Bank," is three hundred and fifty-four miles long and eighty feet high. The greatest deposits, especially of fine clay, are probably made in the quiet depths of the ocean, beyond our observation. By the agency of tidal waves, sand is forced upon beaches; and when aided by storms, the sand is carried beyond the reach of succeeding waves. The isthmus of Suez, between the Mediterranean and Red seas, has doubled its breadth since the time of Herodotus. The collection of loose round water-worn pebbles, accumulated on beaches by waves, is called *shingle.* Very many bays and estuaries, are shoaling rapidly with the sediment conveyed into them by the waves of the ocean. The amount of matter brought down by rivers and deposited as deltas, is quite insignificant when compared with ocean deposits. So great is the quantity of detritus held in suspension by sea-water, that extensive tracts of land which have been purposely flooded with it repeatedly, have been raised five or six feet. The ocean also disperses through its vast extent, the saline substances, as common salt and carbonate of lime, which it obtains either directly from the rocks, or from the rivers which empty into it.

45. The general *result* of atmospheric and aqueous agencies, is the reduction of elevated portions of the earth to lower levels. The loss of land through their influence

greatly exceeds all deposits in the form of dry land; consequently a large portion of the detritus is spread over the bottom of lakes and of the ocean.

IGNEOUS AGENCIES.

46. We have seen that the tendency of atmospheric and aqueous agencies is to destroy the inequalities of the Earth's surface, and deposit the materials thus separated at the mouths of rivers and on the bottoms of lakes and seas. An antagonist force, however—the *igneous agency*—is as constantly operating to restore and produce these inequalities. This agency exhibits itself in the phenomena of *volcanoes, earthquakes, hot-springs,* and *gradual elevations* of extensive lines of coasts and continents.

47. *Volcanoes* are openings in the earth, through which melted rock, lava, smoke, ashes, gases or vapors are discharged. They are usually inverted cones or *craters* at the summits of conical hills or mountains, which vary in height from the smallest hill to nineteen thousand feet, (Cotopaxi.) When they exist upon land they are called *subaerial;* when under the sea, *submarine.* Those which exhibit no evidence of action since the commencement of the historic period, are deemed *extinct.* Of the *active* volcanoes, some are *constant,* others *intermittent.* The periods of intermission vary from a few months to centuries. Monte Epomeo, in Ischia, aft remaining dormant one thousand seven hundred years, again burst forth in the early part of the fourteenth century of the Christian era. Volcanic vents, which emit only sulphurous, watery, and acid vapors, are called *solfataras.*

48. The number of active volcanoes is about three

hundred; of which two-thirds are on islands of the ocean, and the others are usually near the sea. Some, however, are remote from large bodies of water, as Peschan in Central Asia, which is twelve hundred miles, and some of the Mexican and South American volcanoes, one hundred miles distant from the sea. Volcanoes are grouped as *central,* in which the disturbing power manifests itself in radii from a central point, as the Peak of Teneriffe and the Isle of Palma (one of the Canary islands); or *volcanic chains* or *bands,* in which a number of vents occur in a line extending over many miles, oftentimes coinciding with mountain chains. Such a band is presented by the volcanoes of America, those of the Andes and Rocky Mountains being connected by the Cordilleras of Mexico. The same chain may also connect with the remarkable line of volcanoes passing through the Aleutian Islands, Kamschatka, Japan, and the Molucca islands; thence by the Antartic Land (on which Captain Ross observed active volcanoes) to Terra del Fuego, thus encircling the globe.

49. The phenomena of an *eruption* usually commence with rumbling sounds in the earth and emission of smoke, sulphurous and acid gases from the mountain; stones and ashes are thrown with violent explosions from the crater, the earthquake increasing, until the molten lava flows freely down the mountain's sides. Toward the close of the eruption, cinders, red hot stones, and smoke are again thrown out. Impetuous showers of rain, with vivid lightning, and if the mountain is snow-clad, the sudden melting of snow and ice, render the scene still more complicate and awful. The lava sometimes does not rise to the brim of the crater, but bursts through the sides of the mountain and flows over the surrounding country.

50. The greatest exhibition of eruptive violence on record occurred in the island of Sumbawa (one of the Sunda group), in the year 1815. It commenced on the fifth of April, increased in violence until the twelfth, and ceased in July. The explosions were heard in Sumatra, nine hundred and seventy miles west, and at Ternato, seven hundred and twenty miles east of the island. The ashes were carried three hundred miles in the direction of Java, and two hundred and seventeen miles northward toward Celebes in sufficient quantity to produce darkness equal to what is ever witnessed in the darkest night. The floating cinders westward of Sumatra formed a mass two feet thick, several miles in extent, through which ships with difficulty forced their way. Several streams of lava, issuing from the crater of the Tomboro mountains, covered extensive tracts of land and ran into the sea. The area convulsed by this volcanic paroxysm was one thousand miles in circumference. Out of a population of twelve thousand on the island, twenty-six only survived.*

51. Of all eruptions of modern times, the most remarkable in respect to the quantity of lava ejected was that of Skaptaa Jokul, in Iceland, in the year 1783. On the eighth of June of that year, clouds of smoke began to collect in the mountain, obscuring the light of day, showering down great quantities of ashes and sand. On the tenth, slight shocks of earthquake and flames were perceived. On the eleventh, the large river Skaptaa, which had been much swollen, entirely disappeared, and the next day a current of lava rushed down the mountain and overflowed the channel of the river, which was in some places from four hundred to six hundred feet deep, and

* Lyell's Principles.

two hundred feet broad. The lava continued to flow until the twentieth of July, pouring over a lofty cataract, and filling up in a few days an enormous cavity which the river had been for ages hollowing out. On the night of the ninth of August, another torrent overflowed the country to the extent of more than four miles. The eruptions continued, with intervals, till the end of August, and closed with a violent earthquake. One of the streams of lava was fifty miles long by twelve broad, and the other forty miles long by seven broad. Their thickness was, in the narrow channels, five or six hundred feet; but on the plains rarely more than one hundred feet, and in some places only ten feet. Taking the lowest average thickness, the mass of lava can not have been less than twenty thousand millions of cubic yards. Thirteen hundred human beings lost their lives, more than one hundred and fifty thousand domestic animals were destroyed, the fisheries on the coast were ruined, and it is affirmed that Iceland has not yet recovered from the ravages of this eruption of Skaptaa Jokul. The great mass of lava spread over the land by this eruption, distended by gases, cooling on the surface and becoming solid, while the central parts continued liquid and flowing onward, left caverns of great extent and singular appearance.

The cavern of Surtsheller, or the "black cavern," is a long, winding canal, with several branches, enclosed by a crust of lava six feet thick. It is twenty-five feet wide, and its sides and vaulted roof are studded with stalactites of lava and ice.

52. The history of Vesuvius and Etna is more complete than that of any other center of volcanic action, because their phenomena have been for a longer time intelli-

Fig. 19.

Surtsheller Cave in Lava.

gently observed and recorded. The southern part of Italy has, from the earliest periods of human observation, been subject to violent volcanic action. But previous to the Christian era, no record or tradition of eruptions from Vesuvius existed. The summit of the cone, called Somma, as encircled with vines, and its sides were covered with luxuriant vegetation. At its base were large towns, among which were Herculaneum and Pompeii. At this period, Ischia, Procida, and the Phlegrean fields were the scenes of volcanic eruptions.

53. After a slumber of ages, Vesuvius, in A. D. 63, exhibited signs of internal agitation in earthquakes, which increased in frequency and energy until the year 79, when an eruption overwhelmed the cities Herculaneum, Pompeii and Stabiæ. During sixteen hundred years these cities were buried from human observation and memory beneath volcanic ashes and lava. In A. D. 1713, in sinking a well, some pieces of marble and statuary were discovered. This led to extensive excavations in Pompeii and Herculaneum.

Fig. 20.

Vesuvius, showing the site of Pompeii and the River Sarno.

As the former city was buried in ashes and mud, its exhumation is comparatively easy, and has furnished an immense store of antiquities—paintings and sculptures, linen cloth, household utensils, medicines, loaves of bread with the legible stamp of the baker, papyri, &c., in a perfect state of preservation. Few human skeletons have been found, most of the inhabitants having fled before the irruption. Since that period there have been forty eruptions of Vesuvius, some of which have destroyed towns—as that of 1631 in which lava currents with floods of mud overwhelmed Resina which was built over Herculaneum, and that of A. D. 1784, in which Torre del Greco was encased in lava. The eruption of A. D. 1822, was characterized by violent explosions

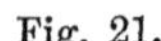
Fig. 21.

Crater of Vesuvius, in A. D. 1829.

which threw out the lava which had consolidated, altering the shape of the crater, and reducing the height of the mountain from four thousand two hundred to three thousand four hundred feet. The preceding cut presents the appearance of the crater in 1829. The amount of lava ejected by Vesuvius in the eruption of 1794, exceeded twenty-two millions of cubic yards.

54. Etna has been known as an active volcano from the earliest periods of tradition. Its height is ten thousand eight hundred and seventy feet. The base of its cone is eighty-seven miles in circumference, and there are numerous subordinate cones or secondary volcanoes upon the sides of the mountain. More than eighty eruptions of this mountain are recorded, several of which are characterized by the great amount of lava ejected; the whole quantity erupted far exceeding the mass of the mountain. The lava of the eruption of 1669 overwhelmed fourteen towns and villages, with a portion of the city of Catania, running into the sea, and covering eighty-four square miles. Its eruptions sometimes, instead of occurring at its summit, take place through fissures in its sides, which abound with lava dikes.

55. The volcanoes of America are generally distinguished by their great height and number. Within two degrees of latitude from the Equator on either side are nine active volcanic vents, including Cotopaxi, Tunguragua, Antisana and Pichincha, all of which are from sixteen thousand to nineteen thousand feet above the level of the ocean. The whole plain on which Quito stands, nine thousand five hundred feet above the sea, with the adjacent mountains, seems to constitute an immense volcanic dome, embracing six hundred square miles, which is almost constantly agi-

tated by internal convulsions, finding vent by eruption at the various craters. Cotopaxi, which rises to the height of eighteen thousand eight hundred and eighty feet, had five great eruptions during the last century, which were heard at distances of from one hundred and forty to six hundred miles on the Pacific coast. Molten lava is rarely raised so high as the summit of these volcanoes, but cinders

Fig. 22.

Cotopaxi.

and pumice are ejected. The most destructive effects are produced by the torrents of mud and boiling water, which result from the melting of the snow with which the mountain is covered, and the bursting of lakes and subterranean cavities. These currents of mud and water fill the valleys to the depth of several hundred feet, and oftentimes contain so many fishes, that their putrescence renders the atmosphere unwholesome for many miles.

56. In Mexico the transverse band has five active volcanoes, of which Popocatapetl and Jorullo have attracted most attention. *Popocatapetl* (Smoking Mountain) attains the height of seventeen thousand seven hundred and twenty feet, is snow clad, and continually emits smoke and vapors.

57. The formation of the volcano *Jorullo* exhibits an

Fig. 23.

Popocatapetl.

instance of the origin of a volcanic mountain suddenly elevated one thousand six hundred and ninety-five feet above the plain, (four thousand two hundred and sixty-five feet above the ocean,) more than one hundred miles distant from the sea, and quite remote from any active volcano. It occurred on the plain of Malpais, west of the city of Mexico, in the year 1759. In the month of June of that year, alarming subterranean noises, with frequent earthquakes commenced and continued about fifty days. After a period of apparent tranquillity, on the night of the 28th of September the inhabitants were compelled to leave the convulsed plain, in which a tract of four square miles heaving like an agitated sea, was raised to a height five hundred and twenty-four feet; flames were seen to issue, and fragments of burning rock were thrown to great heights. Two rivers were precipitated into the chasms, increasing the fury of the flames. Thousands of small cones rose up on the plain from six to ten feet high, called by the Indians, *ovens* (hornitos) emitting sulphureous vapors and smoke. In the midst of these *fumaroles*, stood six large conical masses rising from three hundred to one thousand six hundred feet above the plain. The largest of these was Jorullo which continued burning and throwing up immense quantities of lava containing fragments of granite rock.

The annexed diagram presents the outline of the elevated plain; *a* is the cone Jorullo, and *b* the plain sloping from the base of the cones at an angle of 6°.

Fig. 24.

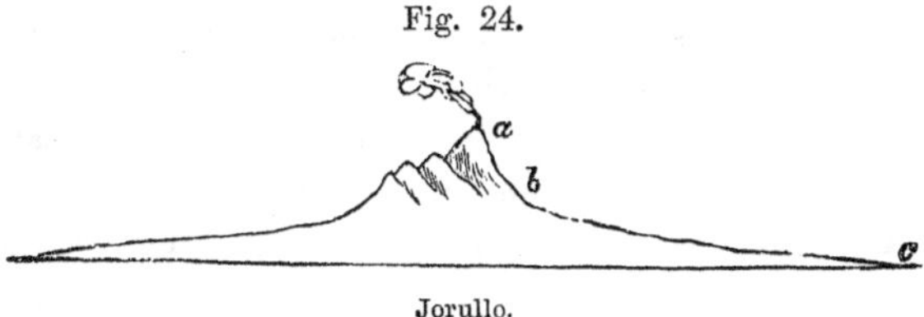

Jorullo.

58. *Kilauea*, in Hawaii, Sandwich Islands, is an ever active, and most remarkable volcano. It is situated, not on the summit, but on the south-eastern flank of Mount Loa, at an elevation of three thousand nine hundred and seventy feet above the sea, while the summit is nine thousand seven hundred and ninety feet higher. Instead of slender walls around a deep crater, liable in most conical craters to be demolished by the explosions of an eruption, the summit of this volcano is nearly a plain, and its crater a deep abrupt pit, seven and a half miles in circuit, of an oval figure, embracing about four square miles. At a depth of six hundred and fifty feet below the brim of this pit, a narrow plain of hardened lava, called the "black ledge," projects like a vast terrace or gallery around the whole interior; and within this gallery, below another similar precipice of three hundred and forty feet, lies the bottom, a vast plain of volcanic rock, more than two miles in length. In this plain are pools of boiling lava, which vary at different times in num-

ber and extent, one of which, in November, 1840, was fifteen hundred feet long by one thousand broad. The lava boils in these pools with various degrees of energy, sometimes flowing over, cooling, and thus raising a rim, or passing in glowing streams to distant parts of the crater. At times, the activity of the ebullition is such as to throw up jets of the lava thirty or forty feet high. The overflowing of the pools raises cones, sometimes one hundred feet high, with central cavities, from which vapors issue.

The adjoining figure represents a singular spire of lava, resembling a petrified fountain. From small vents, the liquid lava, thrown up in jets, falls over, raising a conical

Fig. 25.

Lava Spire at Kilauea.

base. The column is built up by the successive drops of lava which fall upon each other, as they are tossed up. These spires vary in height from a few inches to forty feet, and are miniature craters of eruption.*

59. Four eruptions of Kilauea have occurred since A. D. 1789. That of 1840 was most extraordinary. For several years previous, the lava had been rising in the crater, until it stood about fifty feet above the "black ledge." The immense pressure of the lava and gases opened fissures in the sides of the crater, and the molten flood flowing in a subterranean channel six miles, emerged in an ancient wooded crater, rising in it to the height of three hundred feet; from this again, passing sometimes under ground for miles, and then upon the surface, filling the valleys, melting the hills, and consuming the forests along its path, it poured itself for three weeks with loud detonations into the sea. The length of the stream from Kilauea to the ocean was about forty miles: it accomplished the passage in three days. The breadth of the stream varied from one to four miles, and its depth from ten to two hundred feet. The coast was extended by it a quarter of a mile into the ocean. The whole area covered by it is estimated at fifteen square miles, and the amount of lava at six thousand millions of cubic feet. The waters of the ocean were so heated that the shores were covered for twenty miles with dead fish. Night was converted into day, its glare being visible more than one hundred miles at sea, and at the distance of forty miles fine print could be read at midnight. The lava in Kilauea fell four hundred

* Prof. Dana, in the "Geology of the United States Exploring Expedition"

feet, showing that the eruption was a disgorgement of the lava of that crater. The Sandwich Islands present remarkable regions of volcanic action, in which several craters occur within a few miles.

Fig. 26.

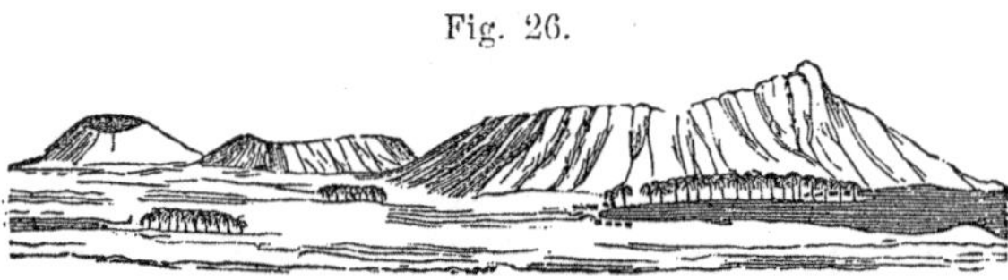

Diamond Hill and adjacent Volcanic Cones in Oahu, Sandwich Islands.

60. The phenomena of *submarine volcanoes* are influenced by the pressure of the water of the ocean, which may at certain depths entirely suppress any exhibition of volcanic agency exerted at the bottom. History abounds with authentic instances of the rise of islands. The Grecian Archipelago is studded with them, and the origin of many of the Aleutian islands is within recorded observation. The structure of others (some of them large, as Hawaii, which covers four thousand square miles, and whose summit is nearly fourteen thousand feet above the ocean,) reveals their history. Teneriffe, St. Helena, and the Azores have the same origin. In the year 1811, an island (Sabrina) arose out of the ocean, near the Azores, to the height of three hundred feet, with a circumference of one mile, and after remaining six months, disappeared.

A remarkable volcanic island appeared in the Mediterranean sea, in the month of July, A. D. 1831, which remained visible above the water about three months. A fortnight previously, shocks were experienced on board a ship passing the spot, which produced an impression like the striking of the ship on a sand bar. On the tenth of July a column of water, like a water spout, was seen rising out of the sea, and soon after a dense mass of steam ascended to the height of one thousand eight hundred feet. On the eighteenth of July, a small island had appeared, with a crater in its center, ejecting volcanic matter. It had, on the fourth of August, attained an elevation of two hundred feet, and a circumference of three miles. After this, it diminished by subsidence and the action of the waves, so that at the end of October, no vestige of the crater remained. In A. D. 1833, a submerged reef, about three-fifths of a mile in extent, existed in its stead.

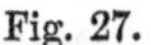

Fig. 27.

Graham's Island, as it appeared in August, A. D. 1831.

61. The term *lava* is applied to any mineral substance which has flowed from a volcano in a melted state. Lavas consist essentially of two minerals, feldspar and augite. When the feldspar predominates, the lava is

called *feldspathic;* it is light colored, and has specific gravity not exceeding 2.8. *Trachyte* is such a lava. When the augite prevails, the lava is dark colored, has specific gravity exceeding 2.8, and is called *augitic*—basaltic. Lava cooled under great pressure is dense, like the older rocks; when cooled under pressure of the atmosphere only, it is porous, distended by gases. Feldspathic lava, flowing into water, is converted into *pumice*, which is so light as to float on the surface of the water. When silex enters largely into the composition of lava, it produces volcanic glass—*obsidian*, which resembles ordinary glass, and is of a smoky hue. A portion of the lava of Kilauea is vitreous, and is sometimes blown by the wind into minute threads, called by the natives "Pele's hair." The lava of this volcano is more fluid than that of most volcanoes. Lava is a very poor conductor of heat, and consequently the interior portions of a lava stream retain their heat a great length of time. Lava ejected from Etna in 1819 was sufficiently hot and fluid to move a yard a day nine months after eruption. In another instance, lava was in motion ten years after it was ejected. The lava of Kilauea, erupted in June, 1840, was so hot in November that pieces of paper introduced into fissures in it were immediately inflamed. Lava flows within a crust which is rapidly formed over its surface. On piercing this crust, the fluid within flows out, and its course may in this way sometimes be controlled. In the summer of A. D. 1828, a mass of ice was discovered on Etna, beneath a bed of volcanic ashes and lava, whose non-conducting property had preserved it for centuries from melting.

63. *Earthquakes* are movements, more or less violent, of the superficial crust of the earth, consisting usually of

rapidly succeeding undulations, oftentimes accompanied by sounds, and traceable in particular directions. Three distinct kinds of motion are recognised. 1st, The *perpendicular*, which acts from below upward, like the explosion of a mine. This was witnessed in the destruction of Riobamba, in A. D. 1797, when many of the bodies of the inhabitants were thrown upon a hill several hundred feet high. 2nd, The *horizontal*, which takes place in successive undulations, proceeding in a uniform direction. 3rd, The *rotatory* or *vorticose*, which seems to be due to interferences of undulations, causing a whirling movement of the Earth, by which buildings are twisted round, and parallel rows of trees are displaced without being prostrated. The first movement is the most common and harmless, while the third occurs only in the most disastrous earthquakes. A hollow *sound*, like a mine-explosion, often accompanies an earthquake, which is sometimes heard as distinctly at a great distance from the scene as in its immediate neighborhood. The *progression* of earthquakes is most commonly in a linear direction, with a velocity of twenty or thirty miles in a minute; but sometimes the concussion proceeds from the center of a circle or ellipse, decreasing in force toward the circumference. Their duration is very brief; thousands of lives are sacrificed, and cities and provinces are reduced to ruins in a few seconds. No country is entirely exempt from their visitations, but particular regions are subject to severe, continuous, and extensive concussions, as Central and South-eastern Asia, South America, and Mexico. Slight shocks are so frequent that there is reason to presume that the surface is continually agitated by concussions on some of its points. There are accounts of no less than three thousand four hundred and thirty-two distinct earthquakes which have occurred in Europe since the commencement of the fourth century of the Christian era.

63. The most remarkable earthquake on record is that which destroyed the city of Lisbon, in A. D. 1755. It commenced with a sudden subterranean sound: this was immediately followed by violent shocks, which demolished the greater portion of the buildings, destroying the lives of sixty thousand persons. A large quay, to which hundreds of people had resorted for safety from the falling buildings, was instantaneously engulfed in an unfathomable chasm, from which nothing ever rose to the surface. Vessels were thrown violently aground; the bed of the river was raised to the surface, and immediately afterward the ocean came rolling in, fifty feet higher than usual. The walls of some houses were seen to open from top to bottom more than a quarter of a yard, and close again so accurately as to leave but slight trace of the injury. The movement of this earthquake was undulatory, progressing about twenty miles a minute, agitating a surface four times as large as Europe and nearly one-twelfth of the whole superficies of the globe. The water in the lakes of Scotland rose and fell three feet. The earthquake was perceived at Fahlun in Sweden, Barbadoes, and on Lake Ontario. The sea rose on the West India islands, and a ship one hundred miles west of St. Vincent suffered so severe a shock that the seamen were thrown upon the deck.

64. The great earthquake of Calabria, Southern Italy, in A. D. 1783, was distinguished by the concentration of its violence, heaving the surface like the waves of the sea. Radiating from the town of Oppido as a center, its violence was manifested over an area of five hundred square miles

About two hundred towns and villages were destroyed, and nearly one hundred thousand people perished. The movement was rotary, as shown by the twisting of the stones of the obelisks of St. Bruno, which were turned from six to nine inches from their former position. A chasm, a mile long, one hundred and five feet broad, and thirty feet deep, was formed; and another, three-fourths of a mile long, one hundred and fifty feet broad, and one hundred feet deep, with numerous elevations and depressions: and many disputes arose respecting the ownership of lands which had shifted position. In the year 1819, a large tract of land at the mouth of the Indus, with villages, was submerged, and another tract, called the Ullah Bund, fifty miles long and sixteen broad, was elevated ten feet.

Fig. 28.

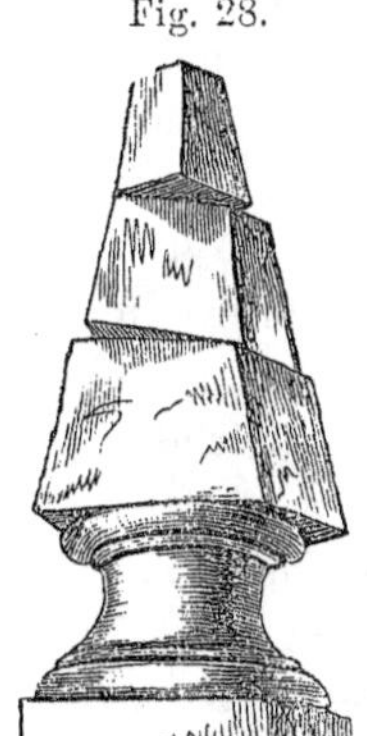

Obelisk of St. Bruno.

65. Violent and extensive earthquakes occur in the vicinity of the Andes, in South America. A terrible convulsion was experienced in 1822 on the coast of Chili, by which an area of one hundred thousa-d square miles was permanently elevated three feet. In 1812, LaGuayra and Caraccas, in South America, were destroyed by an earthquake of great violence. In December of the previous year, a series of convulsions commenced along the valley of the Mississippi, from New Madrid to the mouths of the Ohio and St. Francis rivers. The earth rose in great undulations. Lakes, twenty miles in extent, were suddenly formed, and others were drained. Extensive chasms opened in a direction North-east and South-west, and new islands were formed in

the river. These agitations of the Mississippi valley ceased, when the South American cities were destroyed. This coincidence, together with the direction of the chasms, indicates a subterranean communication between the localities, which are more than two thousand miles apart. The violent earthquake of Guadaloupe, which occurred in A. D. 1842, extended in a direction North-west to South-east, from Charleston, South Carolina, to the mouth of the Amazon river, destroying several towns in the West India islands.

66. *Hot-springs* are common in the immediate vicinity of volcanoes. They are found upon the slopes of Etna and Vesuvius, but in great numbers, and with more imposing features in that remarkable field of igneous agency, Iceland, where they are called Geysers—raging fountains. Within a circuit of two miles more than one hundred of them may be found. They are situated in sight of Mount Hecla, in a plain at the foot of a hill of gray trachyte (ancient lava.) The crater of the great Geyser is a flattened cone of silicious matter; the basin within is of an oval figure fifty-six feet by forty-six, terminating at the bottom in a perpendicular pipe, seventy-eight feet deep. Usually the basin is filled with clear water of the temperature of 180°.

At times a subterranean sound, resembling that made by a volcano during an eruption, is heard, and then a slight tremulous motion is perceived on the rim of the fountain; the surface of the water in the basin becomes convex, and large bubbles of steam rise and burst, throwing up the boiling water several feet high: a heavier noise is heard below, and suddenly there shoots up a column of water to the height of one hundred feet dispersing at the summit into dazzling white foam. After a brief period, a column of steam issues, with a loud, roaring noise; this is followed by another column of water higher than the preceding ones, mingled with stones, accompanied by loud detonations. The phenomenon lasts a few minutes, and then the basin resumes its tranquil state. These waters encrust the surrounding soil, and all substances upon which they fall, with silica.

Fig. 29.

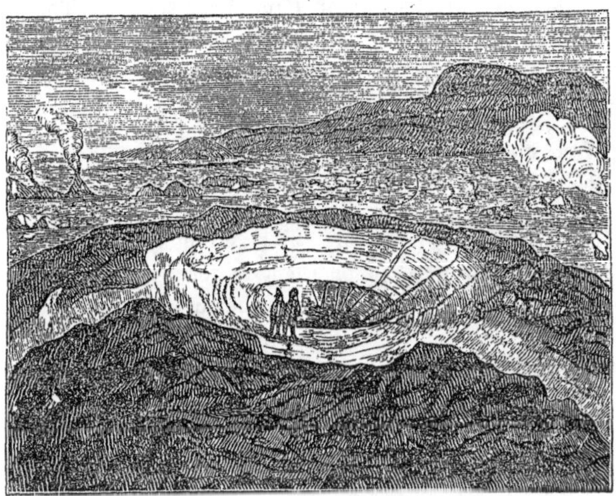

Basin of the great Geyser of Iceland

Boiling springs are found also in the Azores, Java, and the volcanic regions of Central and South America.

ermal springs are not confined to the vicinity of active volcanoes, but are found in the Alps, and Pyrenees, in Virginia, Arkansas, and Oregon. In all cases, however, they are situated near mountains or rocks which have been subjected to igneous agency. The temperature of the water varies in different springs; that of the seventy springs in Arkansas ranges from 118° to 148°. Thermal waters are charged with salts and gases; hence they are mineral waters, as those of Bath in England, Wiesbaden in Germany, Saratoga, and the Sulphur Springs of Virginia.

67. Another apparent manifestation of igneous agency is the *gradual elevation* or *subsidence* of portions of the earth's surface. An interesting example of this is presented by the remains of the temple of Jupiter Serapis, at Puzzuoli, on the shore of the Bay of Baiæ, near Naples. The building was originally of a quadrangular form, twenty feet in diameter, the roof being supported by twenty-four granite columns, and twenty-two of marble: three of them are still standing. The tallest of them is forty-two feet in height; the surface is smooth and uninjured to an elevation of twelve feet, where commences a band of perforations made by the Lithodomus (*lithos*, stone, and *domos*, a house,) a shell-fish inhabiting the Mediterranean sea. These perforations, many of which still contain shells, cover a space of nine feet, and are so numerous and deep as to prove that the pillars were for a long time immersed in sea water; the lower portions were protected by rubbish and puzzolana—volcanic tufa; the upper portions projected above the water, beyond the reach of the lithodomi. The platform of the temple is now one foot below high-water mark, and the sea is one hundred and twenty feet distant. These columns

Fig. 30.

Pillars at Serapis.

have been depressed and again elevated more than twenty feet, the relative level of the land and sea having changed at least twice since the Christian era, so gently that these columns have not been prostrated. Professor Babbage attributes their tranquil depression and elevation to the contraction and expansion of the rocks on which they stand, in consequence of variations of temperature. A small volcano—solfatara—and a hot spring exist in their vicinity. The columns are again gradually subsiding.

68. The *gradual change of relative level* of sea and land, on an extensive scale, in regions remote from active volcanoes and violent earthquakes, is exemplified on the coasts of the Baltic sea and Northern ocean. Beds of marine shells and sunken rocks have been raised above the water-line, and the shells of species now living on the shore are found fifty miles inland and at an elevation, as ascertained by Bravais' measurements, of six hundred feet above the ocean. The northern portion of the coast rises most rapidly; the average elevation is stated to be four feet in a century. *Raised beaches* found in England at a height of from twenty to two hundred feet above the existing sea-level, with shells and all the features of the beaches of the present sea-coast, show the same process on that island. Mr. Darwin has also shown that the southern part of South America, at least twelve hundred miles on the east coast, from Rio de la Plata to the straits of Magellan, and a greater distance on the west coast, has been raised from one to four hundred feet. The north-eastern coast of the United States is supposed to be in the same process of gradual elevation. On the other hand, a large portion of the coast of Greenland has been for four centu-

6

ries gradually sinking. Ancient buildings on low islands and near the coasts, have been submerged. The depression of extensive areas in the Pacific and Indian oceans is proved by the existence of banks of dead coral several hundred feet deeper than the limit at which the animal can have lived. On the coasts of Europe and America are found *submarine forests*, consisting of trees and stumps, together with peat, in the position in which they originally grew, depressed several feet below the level of the sea.

ORGANIC AGENCIES.

69. Although *organic agencies* are less influential in modifying the crust of the globe, than aqueous and igneous agencies, still vegetable and animal bodies not unfrequently make up a large portion of important and extensive rock formations; and are regarded by the geologist with special interest because they furnish the clearest indications of the physical conditions of the globe at the time and place in which they lived.

70. *Marine plants*, which oftentimes cover the surface of the ocean so thickly that ships are impeded in their progress by them, are very perishable, and contribute little to the formation of rocks. But the remains of *terrestrial plants* enter into the composition of soils, and form extensive deposits in the great swamps. *Peat* consists principally of the fibrous roots of mosses, especially of the Sphagnum, which continually throw up new shoots from the decaying extremities below. When dry it consists of from sixty to ninety-nine per cent. of carbonaceous matter, forming a valuable fuel. Peat beds, of from four to twenty feet thickness, are common in Ireland and Scotland. The "moss" on the

river Shannon extends over one hundred and fifty square miles; and one-tenth of the whole island, it is estimated, is covered with peat, called by the Irish, *turf*. It is constantly accumulating, with a rate of increase affected by the amount of moisture and other circumstances; in Europe its increase is estimated at seven feet in thirty years. It is confined to the colder regions of the earth, since the heat of the torrid zone causes very rapid decomposition of organic matter. The process by which it is converted into coal has been, in some instances, observed. In some peat bogs large trees have been found erect; its antiseptic power over vegetable and animal substances is remarkable, preserving them from decay even for centuries. Peat swamps sometimes burst their barriers and deluge the surrounding country with black mud.

71. The floods of large rivers carry down immense quantities of timber, which meeting obstructions accumulate in *rafts*, or pass on to the delta, or the ocean. On the Mississippi and Red rivers, rafts have been formed several miles long, bearing soil and growing trees. The delta of the Mississippi contains many layers of wood undergoing the slow process of conversion into coal; but much of the *drift-wood* passes out to sea, and is conveyed by marine currents to far distant coasts, or becoming water-logged, sinks to the bottom of the ocean. The Icelander is supplied with wood for fuel and building boats, by the ocean, which brings the drift-wood of the Mississippi and the rivers of Central America to the coasts of his island. The vegetable growth of arctic climates is stunted and slow, while that of the tropics is gigantic and rapid.

72. The most efficient organic agency in modifying the crust of the earth, is exerted by the most minute and insig-

nificant members of the Animal Kingdom—the *coral Zoophytes* and *animalcules.*

73. The animals which produce *coral* are very simple, resembling plants both in their figures and colors. Until the last century they were described as marine plants and flowers. They differ from plants in having distinct mouths and cavities to receive and digest food, and sensibility to pleasure and pain. They have no system of vessels for circulation, no glands, no distinction of sex, and no senses but those of touch and taste. Their texture is not that of jelly, but of flesh. They vary in size from a minute fraction of an inch, to eighteen inches in diameter. They live either solitary, or in masses of hundreds of thousands. Individuals are called *polyps;* the whole animal structure, whether simple or compound, is termed a *Zoophyte.* Coral is not a collection of cells in which the polyps may conceal themselves, but an internal skeleton: nor do they exhibit any instinct or industry in forming it. It results from vital processes in their system, which they no more control, than do the more highly organized animals the formation of their bones. The species are perpetuated by eggs and buds. The mode of budding is very similar to the budding of plants. A bud swells and bursts on the side or extremity of the parent, acquires tentacles and visceral cavity, and produces other buds and eggs. Polyps may also be multiplied by artificial section, each part having the power of reconstructing a complete animal.

Every Zoophyte, however large or numerous the colony, commenced as a single polyp; successive budding may have produced myriads of polyps, which eat and digest separately, but all aid in the growth of the common mass. An injury to one of them is felt by the surrounding ones, but not always through the whole mass. Some polyps may be turned inside outward with no apparent injury, and the head of one polyp may be engrafted on the body of another. While the process of budding is advancing at the surface, death is occurring in the central and lower parts of the coral, which, when dead, serve only to support the external living part.

74. Numerous genera and species of these Zoophytes, have been described by Naturalists. A few of them will serve as illustrations of their general appearance.

Fig. 31, presents three branches of the Caryophyllia, the polyps of which are of a bright green color, and reside in the radiating chambers.

Fig. 31.

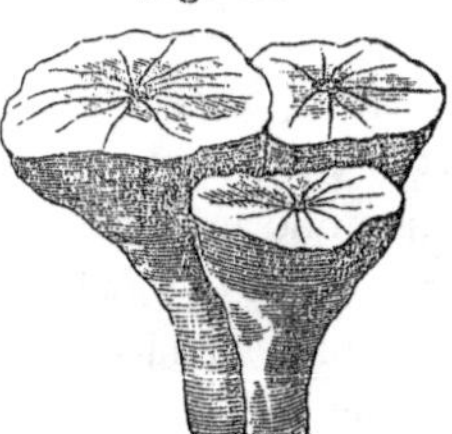

Caryophyllia.

Fig. 32.

Meandrina.

The Meandrina, or brain-stone coral, so called from its resemblance to the convulutions of the brain of animals, as seen in Figure 32, is of a brown color, and attains the size of several feet in circumference. In Figure 33, the same coral appears divested of its fleshy covering, and exhibits the cells within which the polyps partially conceal themselves.

Fig. 33.

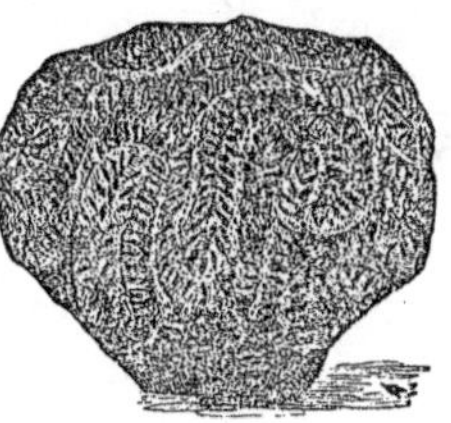

Meandrina without polyps.

The Astrea is a very common and widely diffused species. It

Fig. 34.

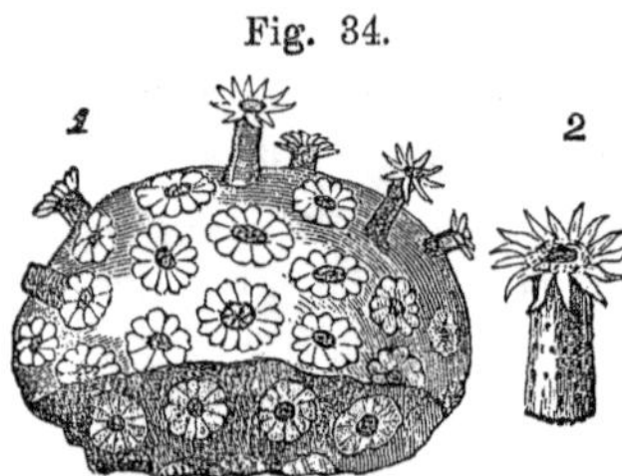

1. Star Coral. 2. The polyp magnified.

derives its name from its radiated or star-like appearance. The polyp is represented in Figure 34, with its tentacles extended. It moves these tentacula, or arms, which are arranged about its head, with great rapidity in taking its food. When the animal is removed, the stellate appearance of the coral is more manifest, as seen in Figure 35.

Fig. 35.

Astrea without polyps.

A still more common genus, seen in cabinets, and as mantel ornaments, is the Madrepore, Figure 36. It is branched and studded throughout with distinct cells.

The Flustra, Figure 37, is a delicate coral, often attached to sea weeds and shells thrown upon the shore. With a microscope, its polyps, if examined in the water, may be seen expanded, and retracted in their cells.

Fig. 36.

Madrepore.

75. Most corals are white, even when the animals secreting them are highly colored. But the Corallium Rubrum—red or precious coral—is of a brilliant red color, while the investing animal is blue. It is obtained from the Mediterranean and Red seas, and is extensively used for ornaments. This species is shown in Fig. 38.

Fig. 37.

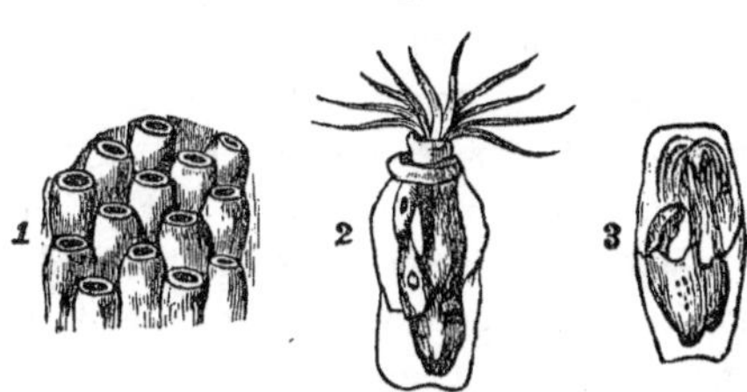

The Flustra. 1. Its cells. 2. The expanded polyp. 3. The polyp in its cell.

An interesting variety of coral is the Tubipora, or organ-pipe coral, (Fig. 39,) composed of parallel tubes, with transverse plates indicating successive generations. It is found growing in the Indian ocean several feet in circumference, and its rich carmine red presents a brilliant ground-work for its polyp of emerald green color.

Fig. 38.

Red Coral.

Ehrenberg speaks in enthusiastic terms of the exquisite beauty of forms, and gorgeous colorings of the corals of the Red sea. "What paradise of flowers," says he "can rival in beauty these living wonders of the ocean."

Fig. 39.

Organ-pipe Coral.

76. It is in the extensive *coral-reefs*, that the Zoophytes evince the power of organic agency in modifying the surface of the earth. The great reef along the line of the

northern coast of New Holland is more than one thousand miles long: a link of three hundred and fifty miles of it is continuous, with no passage or opening through it. Disappointment Islands and Duff's Group are connected by coral reefs so continuous that the natives travel over them from one island to another. Reefs occur in the Pacific ocean from one thousand one hundred to one thousand two hundred miles long, and from three hundred to four hundred miles broad, and of a thickness from thirty to sixty feet, constituting an enormous mass of calcareous matter. These Zoophytes live only in warm seas and near the surface; no indications of them are obtained from deep sea-soundings. Their growth is slow, but incessant, their numbers incalculable: they are usually attached to the shores of rocky islands, or to the crests of submarine ridges, rarely at a depth exceeding sixty feet.

77. Coral reefs are classified as *Fringing*, *Barrier*, and *Circular* reefs: the latter are called, by the natives of the Pacific islands, *atolls*.

Fringing reefs are belts of coral attached to the coasts of islands or continents. When the coast is precipitous, the belt is narrow; but when it is gently sloping, it is covered with coral until it reaches the depth of about sixty feet, where the animals cease to exist.

78. *Barrier* reefs are parallel to the coast, and separated from it by a deep channel. Figure 40 presents the barrier reef of Bolabola, in the Pacific ocean, encircling the island, but separated from it. The reef is in this instance covered with trees. These reefs vary from three to forty miles in diameter. On the ocean side they terminate abruptly in deep water; but within, the slope is gradual. On the outside, the hardier species of Zoophytes

maintain a sturdy growth, resisting a heavy ocean surf, while the frailer varieties flourish in the placid waters within.

Fig. 40.

Barrier Reef of Bolabola.

79. The *circular* reefs, or *atolls,* are the most common forms of coral islands. The diameter of the circles vary from one mile to forty miles, and their breadth from a few yards to more than a mile. They are not always circular;

Fig. 41.

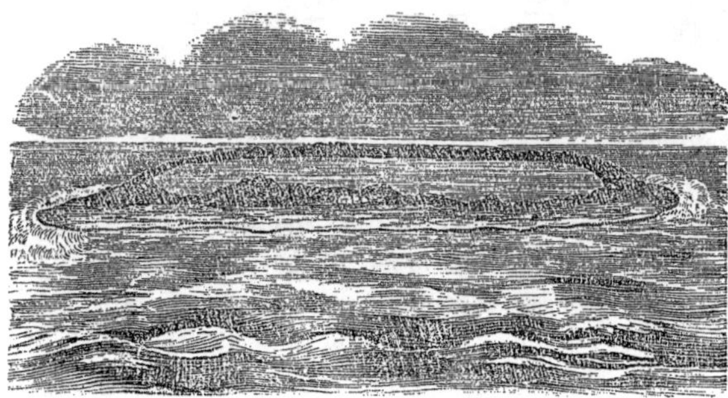

The Coral Isle.—Whitsunday.

one is thirty miles long by six broad. They enclose a space of quiet waters, called a *lagoon,* which communicates with the ocean by one or more openings through the reef. The origin of these circular isles has been the subject of much discussion. They have been supposed to proceed from the growth of the coral upon the circular rims of volcanic craters beneath the surface of the ocean; and some of their phenomena favor such a view. But the subsidence of the islands, about which the corals accumulate as fringing reefs, furnishes a more satisfactory explanation of their origin. Figure 42 presents a section of the island and reef of Bolabola. When the level of the sea was at the lower line, a fringing reef attached itself to the island at A B and B A. As the island sunk in the ocean, the reef grew upward, and formed the barrier reef A′ A′, the upper line now constituting the ocean level. It is now a vertical section of the reef, island,

Fig. 42.

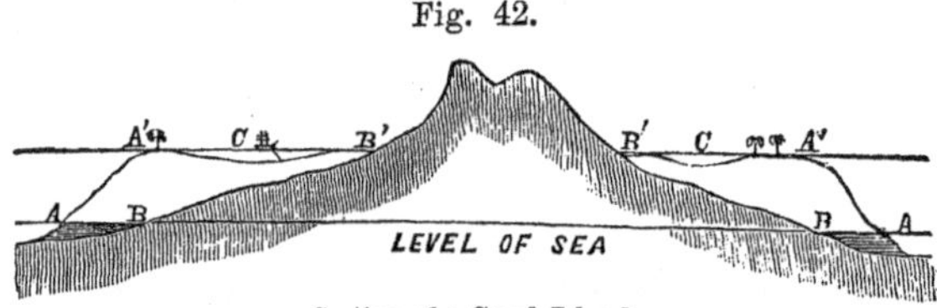

Section of a Coral Island.

and intervening water of Figure 40. When the island has disappeared beneath the ocean level, we have the circular reef or atoll, enclosing a lagoon as in Whitsunday, Fig. 41.

80. But the movement of the islands is not exclusively that of subsidence; many of them have emerged from the ocean, and are still rising. The evidence of this is found in ancient reefs occurring inland and at great elevations above the sea. Upon the summit of the highest mountain in Tahiti, an island composed almost entirely of volcanic rocks, there is a reef of ancient coral attached to the rocks.

This could have grown only in the ocean, and has since been raised to its present position. In the Isle of France also occurs a bed of coral, at a distance from the ocean. A portion of it is enclosed by two streams of lava, in which the characteristic effects of heat are exhibited in its partial crystalization—conversion into compact limestone or marble.

81. Coral reefs, having grown up to the surface of the sea, extend laterally, increasing their breadth. The constant action of the waves accumulates calcareous sand, shells, sea weeds and drift, upon sheltered portions of the reef: in the soil thus formed, seeds, conveyed by the ocean or by birds from islands or the continent, spring up, and the reef becomes a habitable island.

The annexed figure presents a view of the peculiar features of a Pacific island, with its fringing and barrier reefs.

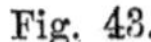

Fig. 43.

A Pacific Ocean Island, with coral reefs.

One of the most singular peculiarities of coral islands is the *shore platform* around them. It is a flat surface, often several hundred feet in width, but little above low tide level. Upon this lie huge masses of reef rock, worn into fantastic shapes. The platform is the result of the abrading action of the sea, and is not strictly confined to coral rocks, but occurs in sandstone shores similarly exposed, as exemplified in Figure 44.*

* Dana's Geology of the United States Exploring Expedition.

Fig. 44.

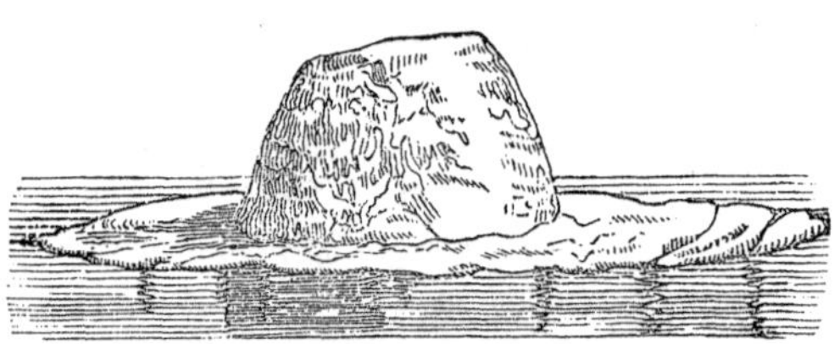

Shore Platform, "The Old Hat," Bay of Islands.

82. The minutest forms of animal life, the microscopic animalcules, commonly called *Infusoria,* from their occurring in great quantities in water infused with vegetable matter, are also important geological agents. Some of them exhibit the simplest conceivable conditions in which animal life can exist; but others show a complex organization, with muscular, nervous and vascular systems. Many of them are covered with shields of silica, or the oxide of iron, whose remains, after the death of the animals, constitute extensive deposits, although they are so minute that 40,000,000,000 of them occupy only one cubic inch. Ehrenberg has demonstrated the existence of monads, which do not exceed the twenty-four thousandth part of an inch in length, and so thickly crowded in the fluid as to leave intervals not greater than their own diameter. Hence he computes that each cubic line of the fluid contains 500,000,000 of these monads. A drop of water, therefore, may include a number of these infusoria nearly or quite equal to the present number of human beings on the globe. They are found in the ocean as well as in fresh water. The beds of marl beneath peat swamps, and at the bottom of ponds, are composed chiefly of their shells, and are often several feet in depth. The red scum seen

floating on the surface of stagnant water consists frequently of the shields of oxide of iron belonging to these animals. Ehrenberg has obtained several pounds weight of the silicious shields of infusoria, which he reared. In case tripoli, or "rotten stone," (which is a mass of fossil infusorial shells,) should become scarce, he proposes to supply the market from this source.

83. The rapidity of their multiplication is most astonishing, one individual of the *Hydatina senta* having been known to increase in ten days to 1,000,000; in eleven days to 4,000,000; and in twelve days to 16,000,000. Of another species, Ehrenberg says, one individual is capable of becoming 170,000,000,000,000 in four days. This rapid multiplication is effected by eggs, buds, and spontaneous division into two or more parts, each one of which very soon becomes a perfect animalcule; this also accounts for their wide diffusion, and sudden appearance in countless numbers. They are found in the waters, upon the land, and in the fluids of living, healthy plants and animals. Only those, however, which have hard shells, leave any trace of their existence after death. Ehrenberg has described about one thousand living species.

Snow is sometimes found in New Shetland and on the Alps, of a *red color*, and occasionally *green*. This is due to an admixture of an infinite number of microscopic plants of low organization—many of them of the tribe of Algæ. They are of globular form, celluler structure, and from one one-thousandth to one three-thousandth of an inch in diameter.

Fig. 45.

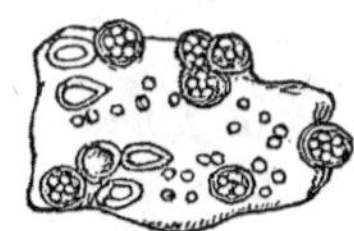

Infusoria in Snow.

Their liquid portions contain myriads of *animalcules.* Fig 45 exhibits a magnified view of these singular beings Their nature is adapted to a very low temperature, so that they can not bear a temperature above the freezing point of water, but when the snow melts they die.

84. Molluscous animals, by furnishing extensive accumulations of *shells,* essentially modify the structure of the surface of the earth. Shell-beds are found beneath the waters, and upon the shores of the ocean, lakes and ponds, producing beds of shell marl. Some shell-beds consist of fragments of numerous species mingled indiscriminately, broken and drifted by the waves and currents; while others are made up of a single species that lived in community, as the oyster, which in some places covers the bottom of the ocean, excluding all other genera, several miles in extent. Some species live only in the mud, while others seek sand for their habitations. Certain species live only attached to the shores, others in shallow water, and others still in water varying in depth from one hundred to one thousand feet. The number of families is much the greatest in shallow water, decreasing as we descend, and ceasing entirely in very deep water. Temperature, nature of the bottom, amount of light and food, determine the residence of each species. After the death of the animals, their shells, protected from decay, constitute in some instances a large part of the mineral bed forming by deposits from the water in which they lived.

85. The remains of *fishes, reptiles, quadrupeds, birds, insects,* and of *men* also, are entombed in deposits now forming, but in much smaller quantities, than those of corals, infusoria and shell-fish. Extraordinary occurrences, as the engulfment of cities by earthquakes, the destruction

of immense shoals of fishes by submarine volcanic agency, the overwhelming of herds of cattle by sudden inundations, and the drowning of clouds of locusts, produce accumulations of their remains, highly indicative of the state of the world at the time in which they lived, just as the fossils of the rocks, formed many ages since, are characteristic of the periods of their formation. But the bodies of most land animals, even the hardest portions of them—their skeletons—undergo decomposition upon dry land, and leave no trace of their forms.

86. The *number* of different kinds—species—of living plants and animals is very great. Eighty thousand species of plants have been described by botanists, and the entire number undoubtedly exceeds one hundred thousand. Professor Agassiz estimates the number of living species of animals at two hundred and fifty thousand; mammalia—those which suckle their young, two thousand; birds, six thousand; reptiles, two thousand; fishes, ten thousand; mollusks—shell-fishes, fifteen thousand; insects and crustaceous animals, one hundred thousand; and the star-fishes, coral polyps, &c., ten thousand. While individuals of each of the species are constantly dying, the species is perpetuated through centuries. Some species, however, have become extinct, within the observation of man. We have no proof of the introduction of a species since the creation of the human race.

87. The *distribution* of plants and animals upon the surface of the earth is very unequal, being influenced by the amount of heat, light, and moisture. Each geographical or climatal region has its own species; which, in the case of plants, constitute its *Flora*, and of animals, its *Fauna*. There are three great climatal regions,—the arctic, temperate

and tropical. The vegetation of the arctic region is confined to mosses, lichens, and a few trees of stunted growth. The Flora of the temperate region embraces the nutritious grains and fruits, with lofty trees of dense fibre, durable and strong —the pine, oak, and cedar. The tropical region greatly excels the others, in the variety and luxuriance of its productions. The plants of different continents in the same latitude are quite unlike; those of Africa, for example, bearing little resemblance to those of South America or New Holland on the same parallels of latitude, each having been created in its own station.

88. The *Faunas* also are capable of distribution into three principal divisions, in accordance with climate, viz., the arctic, the temperate, and the tropical faunas. The plants and animals found at high elevations on mountains within the torrid zone, resemble those of colder latitudes.

The principal characteristic of the *arctic* fauna is its uniformity, embracing few species, but very great numbers of individuals in each species. The same animals are found in it on the three continents, America, Europe and Asia. Some large quadrupeds belong to this fauna, as the moose, the white bear, reindeer and musk-ox. Whales and seals abound in the polar seas, together with star-fishes, jelly-fishes and small crustaceous animals, upon which the whale principally subsists: but very few polyps, and none which secrete coral are found in these seas. Very few insects live in this zone, and no reptiles. The color of the animals of the arctic fauna is frequently white, as shown by the white bear, the white fox, and the ermine, and when of other hues is not brilliant.

While the number of individuals of the *temperate* fauna is no greater than that of the arctic the number of species is much greater and more varied. Very numerous orders and genera of animals, with strong contrasts of form and color, are here presented. The members of this fauna on different continents are similar: some of the families, genera, and a few species are identical. The arctic and temperate fauna are not separated from each other by any very sensible limit, but gradually pass into each other; a few species of animals range through the entire extent of both of them, as the musk-rat, the ermine and the European field-mouse.

The predominant feature of the *tropical* fauna, is its great variety of animals with coverings of brilliant hues. Its members on different continents are quite unlike each other; they are, however, more nearly allied to each other than to the members of the other faunas.

15. Besides faunas separated from each other by difference of climate, we have them more or less distinctly limited by geographical features. The interposition of mountain chains, deserts and seas separate faunas in the same latitude. The animals of the prairies of America, the steppes of Asia, and the deserts of Africa, are peculiar to those localities. The fauna of Oregon and California is said to be more unlike that of New England, than the European fauna is. Marine animals are distributed in the same way, into local faunas. The codfish does not wander far from the Newfoundland Banks. The fishes of the coast of South Carolina are different from those of the West Indies. Faunas that differ much are frequently found near each other, while similar faunas are oftentimes widely separated. The range of a species is not affected by its powers of locomotion. The reindeer is no more apt to transcend the limits assigned it, than is the oyster. The distribution

of animals is not the result of external influences, for were it so, we should always find in the same circumstances similar animals; but it is a law of their being, established by their Creator, analogous to the instincts with which He has endowed them for self-preservation.*

90. Among the causes modifying the structure of the globe, must be recognized the agency of *man*. This agency is exerted in controlling to a certain extent the operation of other agencies, aqueous, igneous, and organic, and in contributing to the rocks now forming peculiar mementoes of his existence. Human agency is apparent in the leveling of portions of the earth's surface; in the direction and restraint of water courses; in the destruction of certain species of plants and animals, and in the substitution of other species. By this means some genera of animals have become extinct, and others are very greatly reduced in numbers. Beasts of prey disappear before advancing civilization, and those animals which are sought for human purposes, without domestication, as the beaver, the seal and the whale, are subject to incessant invasion. Since the discovery of South Georgia, in 1771, one million two hundred thousand seals have been annually destroyed, for their skins; the animal is becoming extinct in that locality. In place of the trees of the forest, man substitutes other plants, particularly the grasses, thus essentially interfering with the natural laws of distribution.

91. Human skeletons are found in rocks recently formed, as in the limestone of the shore of Guadaloupe. Portions of these skeletons are now in the museums of London,

---

* Principles of Zoology, by Agassiz and Gould, chap. xiii.

Paris, and Charleston. They were found, together with stone hatchets, arrows, and pieces of pottery, in a rock consisting of fragments of corals, shells and sand, cemented pretty firmly by the carbonate of lime held in solution in the water. The shells and corals belong to species that now reside in the sea in that vicinity. Figure 46 presents

Fig. 46.

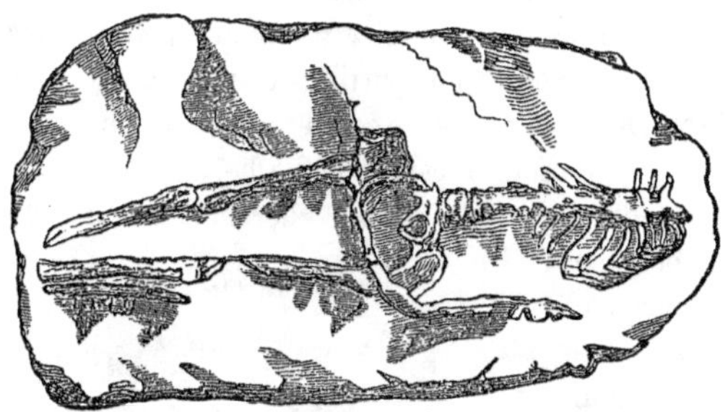

Human Skeleton in the Limestone of Guadaloupe.

a view of the specimen in the British Museum. Other entire skeletons have been disinterred from the same rock; some of them found in a sitting posture. The bones have not been petrified, but contain a portion of their gelatine and the whole of their phosphate of lime. Stone hatchets, and a piece of guaiacum wood having rudely sculptured on one side a mask and on the other a frog, have been found in the same bed. Human bodies clad in skins have also been found preserved in peat swamps in England, with coins, arms, and other implements, such as were used by the Britons at the time of the Roman invasion. Great numbers of human beings have been destroyed by sudden catastrophes, as earthquakes and inundations. Thousands were swallowed up by the great earthquake of Lisbon, (§ 63,) and by that of 1780 in Jamaica, which sent a wave of the sea over the city of Savanna la Mar, sweeping off every inhabitant. In 1787, a hurricane drove the sea upon the coast of Coromandel, twenty miles inland, overwhelming ten thousand of the inhabitants with a deluge of mud.

92. Not only are the bodies of men thus preserved, but the products of their skill, as coins, earthenware and glass. Thousands of ships are annually wrecked upon the ocean, inland seas and lakes: such portions of their cargoes as are not rapidly corroded or decomposed by the water, are soon enveloped by the mud or sand of the bottom, and thus become a part of the rock forming there. In figure 47 we have a specimen of conglomerate of sand, glass beads, knives, &c. cemented together by the oxide of iron which was dug up from the bed of the river Dove, in Derbyshire, England. Two silver pennies of Edward I. are enclosed in the specimen, which are supposed to have been a portion of the treasures of the Earl of Lancaster, lost while crossing the river, more than five centuries ago.*

Fig. 47.

Coins cemented in recent Sandstone.

* Mantell's "Wonders of Geology."

93. An examination of the phenomena of Nature, with reference to the causes now modifying the structure of the globe, exhibits an incessant series of changes, slow and imperceptible, or sudden and conspicuous, by which the features of the Earth are essentially altered. The atmospheric and aqueous agencies continually wear down the dry land, and if not counterbalanced by other forces, would ultimately reduce it to the level of the ocean; while the organic and igneous agencies are accumulating and elevating mineral matter above the ocean's level. What is now dry land was, some thousand years since, the bottom of the ocean, and the present seas cover portions that were then dry land. The geological history of the world is discovered by the study of the deposits thus made, and thus elevated. Some degree of familiarity, therefore, with the operation of these natural causes on a large scale, is indispensable to the geological student, to enable him to interpret the meaning of those past events, the results of which constitute the phenomena of geology.

CHAPTER II.

THE STRUCTURE AND POSITION OF ROCKS.

94. The term *rock* in Geology indicates any aggregation of minerals, hard or soft, compact or loose. The desert of Sahara is a sand rock; so in like manner masses of clay and gravel are rocks. The structure and position of rocks depend upon their origin. The most obvious distinction of mineral masses is into *stratified* and *unstratified* rocks.

Stratified rocks are such as occur in layers included between nearly parallel planes; varying in thickness, from a fraction of an inch to many feet. The whole mass of rock is sometimes called a *stratum*, and the parallel subdivisions of it are termed *beds* or layers; the more minute subdivisions are laminæ, which are generally parallel to the planes of stratification. The term *bed* is also applied to a mass which is wedge-shaped, or lenticular, as a bed of gypsum, salt or coal. Such beds are said to be *subordinate* to the strata in which they occur.

As strata originate from deposition in water, the stratified rocks are termed *aqueous* and *sedimentary*. When the deposit is made upon a level surface and in quiet water, parallel horizontal laminæ are formed; but materials deposited upon a steep shore, produce oblique lamination. When the depositing waters are agitated by waves, the laminæ are waved, and exhibit what are called *ripple marks*, as exemplified in Fig. 48.

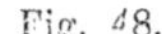

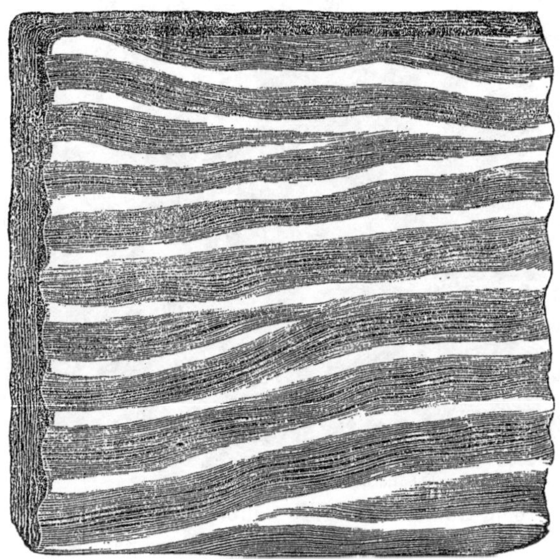

Ripple marks in the New Red Sandstone.

Such ripples may be seen in the sand and mud not only on the shores, but at the bottom of rivers, lakes and the ocean. Laminæ sometimes occur highly curved, and twisted. They could not have been deposited in these shapes, but have assumed them after deposition in consequence of the unequal pressure they have sustained.

95. The stratified rocks consist of *fragments* of crystalline minerals, which are made to cohere by pressure, or by some cement; hence they are called *mechanical*, to dis-

Fig. 49.

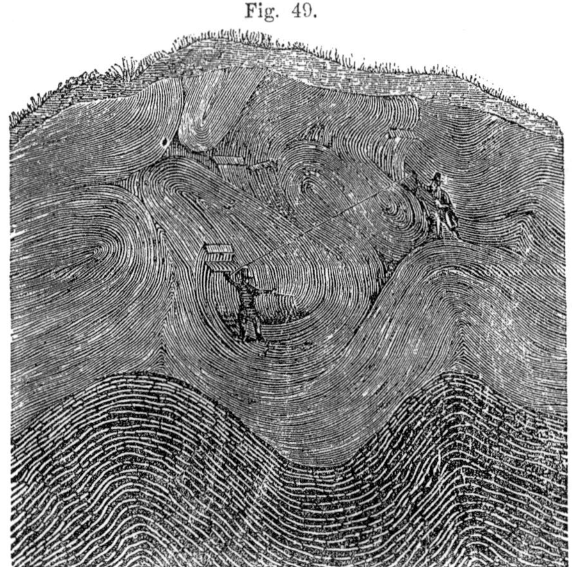

Curved Sandstone in Erie County, Ohio.

tinguish them from those which exhibit a crystalline structure, due to chemical agency. But the subordinate beds of rock-salt, gypsum, &c., are chemical precipitates from solution, and some stratified rocks bear the characters of both agencies.

Fig. 50.

Septarium.

96 A *concretionary* structure oftentimes pervades rocks. The forms of the concretions are various—spherical, ellipsoidal, lenticular, &c. Sometimes, by compressing each other, they become indented and assume various irregular shapes. They are frequently crystalline and concentric, having a leaf, stick, fish, or other organic substance as a nucleus about which they have accumulated. Some of them have fissures within, dividing the mass into irregular shapes, which sometimes resemble the markings of the turtle's shell; hence they are called *turtle-stones.* The name commonly applied to them is *septaria,* (from *septum,* a partition.) The crevices are often filled with calcareous spar—crystallized limestone. From these is prepared an excellent hydraulic cement.

Fig. 51.

Septaria in the banks of the Huron river.

The adjoining cut exhibits the septaria in the slaty banks of the Huron river. Many of them are worn out of the banks and precipitated into the river: some of them are several feet in diameter. Some of the claystones found in England are so regular in figure and so smooth as to have given rise to the supposition that they were turned in a lathe, and to have been used for money. In this country, they are usually thought to be the work of water or of the Aborigines.* They are caused by molecular attraction, and occur mostly in clay rocks. Similar concretions of iron ore occur in regularly ellipsoidal figures in the sandstones associated with the coal, and are called *kidney iron ore.* Concretions are sometimes arranged in layers in a portion of the rock, while other portions are entirely free from them. They sometimes consist of alternating coats of calcareous spar and iron ore.

97. A *concentric structure* is of frequent occurrence in the shales and sandstones of New South Wales. Professor Dana gives a remarkable instance of it, illustrated by the accompanying figure. On either side of a vertical fissure

Fig. 52.

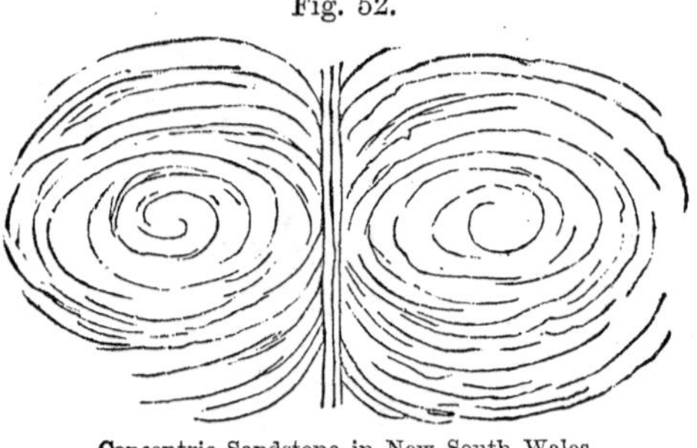

Concentric Sandstone in New South Wales.

is a circular area of ten feet, in which the concentric coats of sandstone are from half an inch to two inches in thick-

* Hitchcock's Geology.

ness. This structure gradually loses the curvature and disappears. Globular concretions also occur here, resembling cannon balls dropped into mud, which are compact and hard, having some foreign bodies, as carbonized wood or pebbles, for their nuclei, though not always at the center of the concretions. In some places they are ten feet in diameter, covering the surface like artificial domes, and looking like a village of rounded huts.*

98. With reference to their mineral ingredients most stratified rocks are included in the three following divisions—the *sand group*, the *clay group*, and the *lime group*. The members of these groups exhibit various degrees of fineness and compactness in their structure, and are designated *arenaceous*, *argillaceous*, and *calcareous*, as sand, clay, or lime is the characteristic ingredient.

99. Stratification is the most general condition of the rocks constituting the crust of the earth, covering nine tenths of its surface. Stratified rocks always overlie each other in a constant *order of succession*. A stratum, which in any one situation underlies another, will never, in any other situation, be found above it. Certain strata may be in some places deficient, but all those which occur together are invariably in the same relative positions. Thus if six strata be designated by the letters A, B, C, D, E, F, in the order in which they succeed each other, and B, D, be deficient in any locality, the order of the others will always be A, C, E, F. In some instances the strata have been displaced so as to bring them in an order of succession different from that in which they were deposited, as is shown in Fig. 53. where

* Dana's "Geology of the United States' Exploring Expedition."

the strata have been folded in such a way as to cause a repetition of them in an inverse order. Here stratum 6 was the lowest, and the others rested upon it in the ascending order 5, 4, 3, 2, 1. The upper part of the curvature has been worn off, so that the strata appear at the surface *a*, *a*, in an unusual order of superposition. Such an instance is called a folded axis.

Fig. 53.

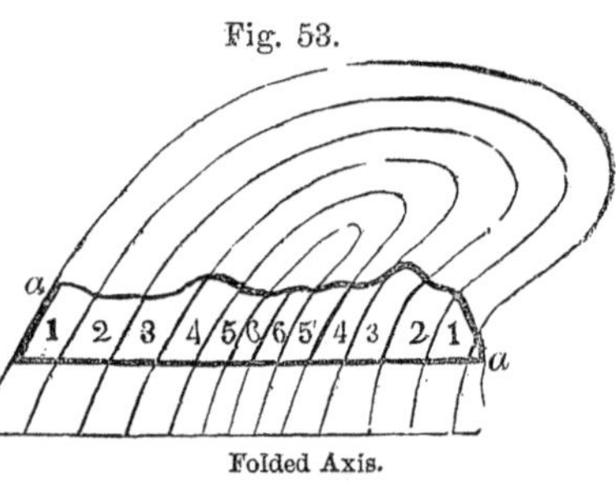

Folded Axis.

100. Strata are deposited horizontally in obedience to gravity and some of them retain the horizontal *position*, but most strata are inclined to the horizon, having been elevated to various angles, by subterranean forces, since their deposition. In Fig. 55, we have four strata deposited horizontally. Figure 56 presents the same strata elevated by subterranean forces, with the uppermost ones rent. Subsequently atmospheric and aqueous agencies wear off the upper

Fig. 54.

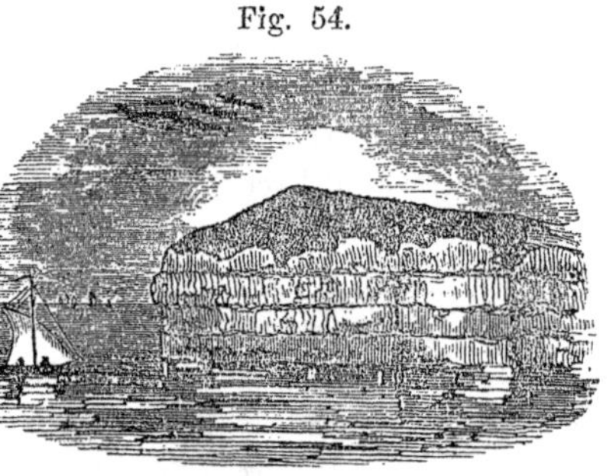
Horizontal Strata.

Fig. 55.

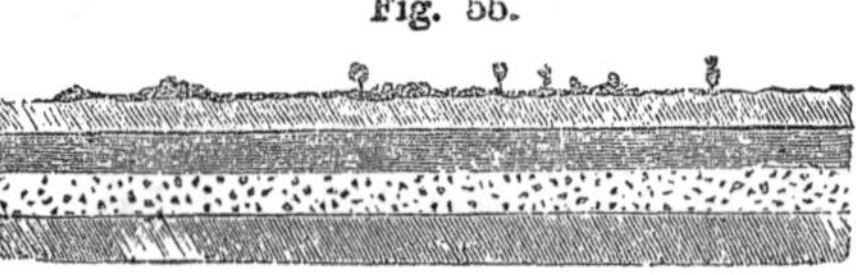

Fig. 56.

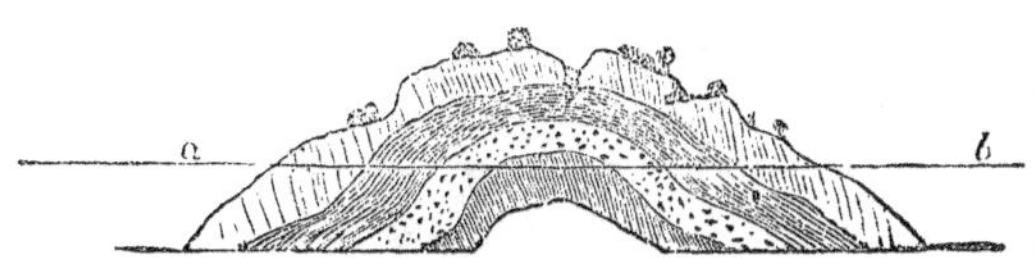

portion of the bent strata, which now appear inclined to the horizon as in Fig. 57.

Fig. 57.

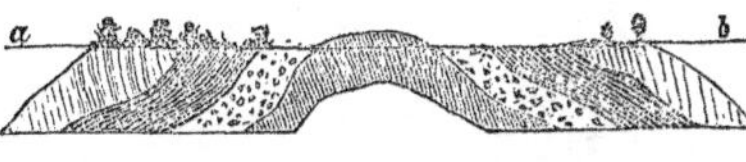

The *dip* of a rock is the angle which the plane of the stratum makes with the plane of the horizon; and is reckoned from 0° to 90°. When the dip is 90° the strata are, of course, vertical. In the Isle of Wight is a series of strata eleven hundred feet thick in this position. The example in Fig. 58—strata in Wales, on which Powis Castle is built, is given by Mr. Murchison, in his "Silurian System." Vertical strata occur on a much larger scale in the cliffs of Savoy. Strata of calcareous shale in the Alps stand vertical for more than one thousand feet in depth and then curve round to their appropriate position.

Fig. 58.

Vertical Strata.

101. The dip of the strata is determined accurately by means of an instrument called a clinometer; but it may ordinarily be estimated with sufficient accuracy by the eye. As a general fact, the deepest strata are most highly inclined. The direction in which the edge of an upturned stratum appears at the surface is called the *strike* or *bearing*. If a ridge runs North and South, the dip of the beds is East and West, and their strike or bearing North

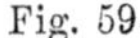

Fig. 59.

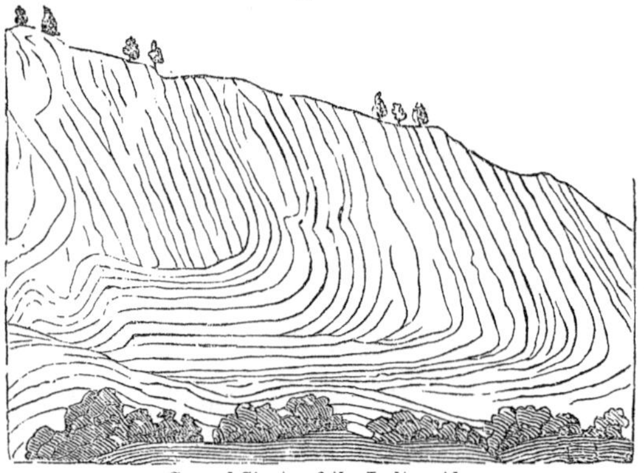

Curved Strata of the Iselten Alp.

and South. A pocket compass will enable the observer to determine the strike. The line of dip is always at right angles to that of strike. Horizontal strata have neither dip nor bearing. If we place a book upon the table with the edges of the leaves downward, as in Figure 60, and remove one cover a short distance from the leaves, this cover may represent a dipping stratum, the dip becoming less and less as the cover is raised, until it becomes horizontal—parallel with the table—when the dip ceases.

Fig. 60.

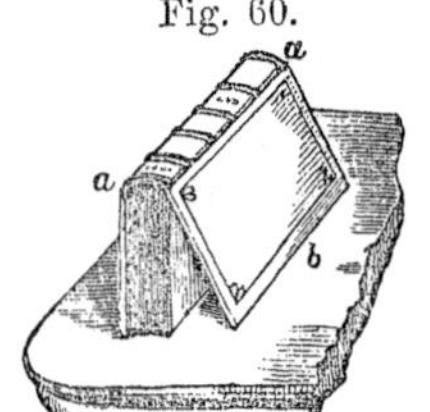

The back of the book, *a a*, exemplifies the strike. The *anticlinal* line or axis is a line along the summit of a ridge or mountain range, from which the strata dip in opposite directions. If both covers of the book be thrown partially open, the anticlinal axis will be represented by a line along the back of the book. The *synclinal* line or axis is the line in a valley toward which the strata dip. To represent this, turn the book over, placing the back upon the table, open it partially and the line between the pages will present the synclinal line.

102. The dip is usually easily discerned, but as the edges of highly inclined strata may give rise to horizontal lines on the face of a vertical cliff, as seen by an observer in the line of their strike, their dip would not be apparent. A break in the cliff, giving a section of the strata at right angles to their strike, would at once discover their dip. Thus the strata in the headland, Figure 61, would appear perfectly horizontal to an observer in the boat directly in front, while a person on the shore facing a section at right angles to the strike of the strata, would at once perceive that they dip 40°. The abrupt termination of strata in a

Fig. 61.

Apparent horizontality of inclined Strata.

headland is called an *escarpment*. When the strata dip in all directions from a point, as around a crater of a volcano, the line of strike is circular or elliptical, and the dip is said to be *qua-qua-versal*. When the strata come out at surface they are said to *outcrop*.

103. An assemblage of rocks formed under the same circumstances, consequently possessing some characteristics in common, is called a *formation*. It often embraces different substances; the Lias formation includes the Lias limestones, shales and marls, as does the Coal formation the rocks associated with the coal. The time during which such a group was formed is called a geological *period*. When successive strata or groups of strata are parallel to each other, they are said to be *conformable;* when not parallel, they are *unconformable*. In Figure 62, the strata *a b c d* are conformable, as are also *e f g h;* but the two groups or formations are unconformable. This indicates

Fig. 62.

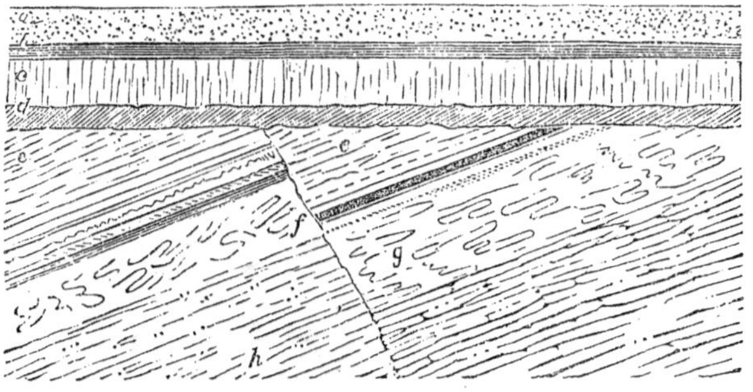

Conformable and unconformable Strata.

that the group *e f g h* had been formed and elevated before the other group was deposited upon them. As the stratification of different formations is usually unconformable, it is inferred that there have been several different periods during which the various formations were deposited and elevated. The elevation of the strata has not always been perfectly equable; hence fissures occur, on either side of which portions of the same stratum are found at different heights, as at *f* in Figure 62. These interruptions of the strata are particularly troublesome to miners working beds of coal or ore, and hence they have been called *troubles, faults,* or *slips.* The fissures are usually filled with sand, earth, and angular fragments of rocks. When the fissure extends to the surface and has considerable width, it is termed a *gorge;* when it is still wider, it is called a *valley.*

104. The *thickness* of strata is determined by measurements applied to their edges. If they are vertical, a measure applied horizontally to the edges gives their thickness; but if they are inclined, it is ascertained by a simple trigonometrical process. Having measured the breadth of the upturned edge, and ascertained the dip, we have the hypothenuse and angles of a right-angled triangle, from which the perpendicular side—the thickness, is easily obtained. The total thickness of the strata is various in different places. Dr. Buckland estimates the thickness of European strata at ten miles. The stratified rocks usually contain remains of plants and animals, and are then called *fossiliferous.*

105. The *unstratified* rocks exhibit no arrangement in parallel layers, are of crystalline texture, and, having undergone the action of heat, are called *igneous* rocks. They occur in three different positions: beneath all the stratified rocks, granite being the deepest known rock; above the stratified rocks, constituting the summits of the loftiest mountains; and thrust into the strata, as veins and dikes. The unstratified rocks are found principally in mountains, and are not widely diffused at the surface, of which they constitute not more than one-tenth, but beneath the thin crust of strata are supposed to form the great mass of the globe. They cause extensive changes in the characters

and position of the strata with which they come in contact. The unstratified rocks are entirely devoid of the remains of animal and vegetable bodies.

106. *Veins* are usually masses of igneous rocks injected from below into fissures in both stratified and unstratified rocks; sub-dividing as they advance, and becoming mere

Fig. 63.

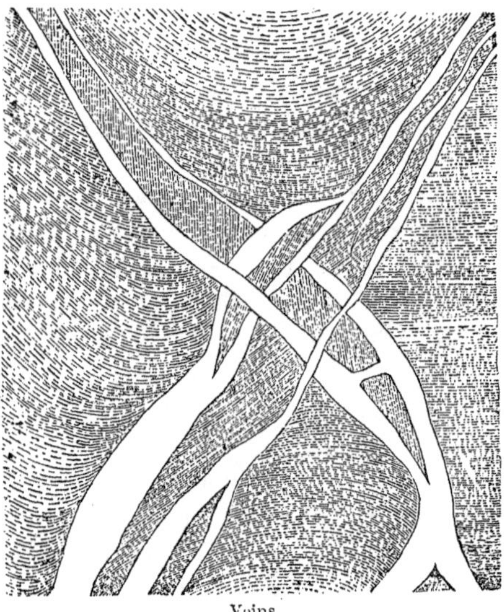

Veins.

threads, they disappear. They are frequently chemically united to the sides of the fissures, but sometimes do not adhere to them. Their contents are sometimes influenced by the characters of the rocks through which they pass. *Dikes* are large veins of trap-rock, porphyry or lava, extending in some instances seventy miles, with a thickness of several yards. Dikes are nearly straight, while veins are very tortuous. As dikes are very compact and hard, the strata through which they pass are often worn away, and they are left standing out like walls. Dikes and veins frequently intersect; in which case that which cuts through the other must be the last erupted, and hence several successive periods of eruption are proved in the granite, trap and other igneous rocks. Some veins are found entirely in-

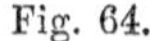

Fig. 64.

Trap Dike.

cluded in the rock, and are not traceable to any mass of similar composition but appear to have separated from the rock in which they lie. These are called veins of *segregation*. Veins and dikes cross the strata at various angles, and are sometimes intruded between the strata, and spread out so as to resemble true beds; but they have no lamination. The contents of veins are exceedingly varied; indeed it is presumed that they contain all elementary substances known to chemists. They are divided into two classes—the *metalliferous* and *non-metalliferous.* The contents of dikes are much more limited in kind and uniform in character.

107. A concretionary structure on a large scale is occasionally seen in igneous rocks, but a more interesting structure exhibited by them is that which produces regular columns, varying in size from an inch in diameter to several feet; in length, from one to three hundred feet; and in number of sides, from three to twelve. They are so accurately adjusted to each other that no space intervenes between them, and frequently consist of joints with alternate convex and concave surfaces: they are usually straight, but sometimes highly curved.

Fig. 65.

Jointed Columns.

108. Mr. Lyell distinguishes the igneous rocks into two classes—the *volcanic,* and the *plutonic.* The volcanic are those which have been produced at or near the surface of the earth, as are the lavas of volcanoes of the present period; but similar rocks have been poured out upon the land or the bed of the ocean, and have been injected into fissures near the surface, at many different epochs. The plutonic rocks appear to have been formed under enormous pressure at great depths in the earth. They differ from

the volcanic in being more highly crystalline, and free from the pores, or cellular cavities which characterize the volcanic. The granites and porphyries belong to this class.

109. There is another class of rocks, which partake of the characters of both aqueous and igneous rocks. These lie upon the plutonic rocks, are highly crystalline in structure, are destitute of organic remains, and yet are divided into beds precisely like the sedimentary formations in form and arrangement. These strata appear to have been deposited in water in the usual way, and then to have been subjected to such a degree of subterranean heat as to assume a new texture. In some instances a portion of a stratum has exchanged its earthy for a crystalline texture through a distance of a quarter of a mile from its contact with granite, and has had all traces of its organic remains obliterated, while the remainder of the stratum retains all the characteristics of its sedimentary origin. Thus dark colored limestones filled with shells and corals have been converted into white statuary marble, and clays into slates, or schists. These altered rocks are called *metamorphic*.

110. All rocks, whether stratified or not, are divided into masses of determinate figures, by natural fissures traversing them in straight lines, and forming planes of variable width. These fissures are called *joints*. Their faces are usually smoother and more regular than the planes of stratification, to which they are vertical, thus dividing the rock into cubical or prismatic blocks. Some joints are more open, regular and continuous than others, passing through several alternations of strata, dividing them from top to bottom, sometimes completely arresting the cross joints. These are called *master-joints*. The constancy of

direction of these fissures is such as to indicate a general and long-continued agency pervading the whole strata. In Great Britain, two-thirds of them run north north-west to south south-east, and the remaining third at right angles to that direction, independently of the dip or strike of the strata. Their origin is attributed to contraction during the consolidation of the strata; to expansion and contraction by alternations of temperature; and to electricity. In some sandstones and beds of ironstone, there are numerous and irregular fractures or seams dividing the surface into small polygonal areas with a concentric structure, as exhibited in Figure 66.

Fig. 66.

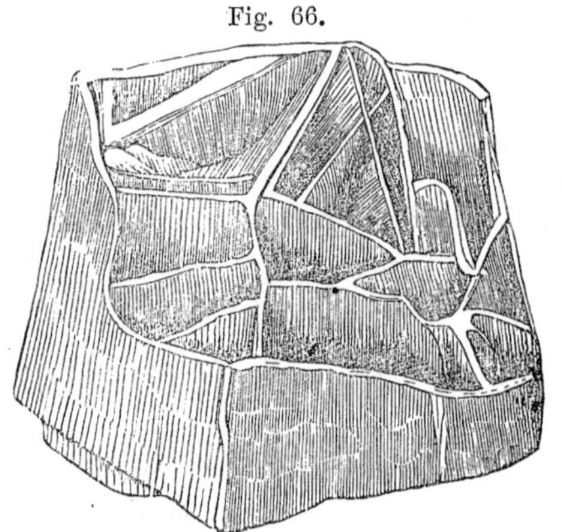

111. Some slate rocks are capable of indefinite subdivision in a direction not coinciding with the planes of stratification, nor with joints. This is termed *cleavage*. The direction of the cleavage planes appears to be, generally, parallel to the anticlinal line of the region in which the rocks occur, and is altogether independent of the dip of the strata. The strata on the two sides of a mountain chain may dip in opposite directions, while the cleavage planes are vertical between them, and parallel to the anticlinal axis, as is the case in the Alps. The phenomena of cleavage in rocks are ascribed to crystallization, or a re-arrangement of the par-

ticles of the strata, by which similar materials are collected in planes.

CLASSIFICATION OF ROCKS.

112. While little difference of opinion exists among geologists with reference to the general arrangement of the rocks, several systems of classification have been proposed by them. These systems differ in the grouping of particular strata and formations. Their diversity is a source of distraction to the student of geology, on account of the multiplicity and discordance of the terms, which they introduce. All geologists agree in the division of rocks into—stratified and unstratified—fossiliferous and non-fossiliferous—and in the invariable order of succession of the stratified fossiliferous; but it is not practicable at present to determine, in geological formations, the relative places of classes, orders, genera and species, with that accuracy which characterises some other branches of natural science.

113. The fundamental idea involved in systems of classification is the relative *age* of rocks, and formations. This, in the case of stratified rocks, is determined by the *position* of the strata; by the characters of the *animal* and *vegetable* bodies they contain; and by their *mineral* constituents.

114. The order of superposition of strata is manifestly indicative of their relative age, since the lowest stratum, upon which the others lie, must have been first deposited, and the others in order upward. But a difficulty attends the investigation of this order of succession, on account of the absence of some of the strata at any one place of observation. There is no place on the globe, where, if a section were made through the rocks, all the strata would be foun

because the strata are not continuous round the earth, like the coats of an onion. Those parts of the globe that constituted the dry land while any deposit was forming in the seas, would receive no portion of the deposit, and as irregularities of distribution of land and water have always existed, parts of the surface must have successively constituted continents and islands. This difficulty in observation is obviated by the meeting and overlapping of the various formations. Thus, if six formations be represented by the first letters of the alphabet *a, b, c, d, e, f,* in their order, and at one place the formations *b* and *e* be missing, observation at another place may give the formations *a, b, c, d, f,* thus supplying the first deficiency *b;* and further investigation at other points, present *c, d, e, f,* thus enabling us to determine the order of succession of the six formations.

115. Since each formation has *fossils,* remains of animal and vegetable bodies, peculiar to itself, we are enabled by means of them to determine whether strata which are remote from each other geographically, as in America and Europe, were deposited at the same or different periods. A difficulty may seem to arise here, from what has been stated (§ 87) respecting the different animals and plants which live at present in different localities, but the differences between the faunas and floras of different geological periods are much greater than those which exist between the animals and plants of any one period. The difficulty, in the case of the older formations, is also much diminished by the great uniformity which characterised the faunas and floras of those periods; they were less numerous and more widely extended than at the present period. The relative ages of the more recent formations may be determined to a certain degree, by observing the number of animals and

plants contained in them which are identical with species living at the present time. This number continually diminishes as we recede from the present geological period, until no trace of the species that live at the present day is found. When two geological formations contain many fossils in common, we infer that they were formed at about the same period.

116. Some *minerals* have been deposited, at certain periods more abundantly than at others; still, different minerals have been deposited at the same period, their distribution depending upon local circumstances, as at the bottom of the present ocean, in some places limestones are forming, at others clay-beds, and at others sandstones. Identity of mineral constitution, therefore, does not prove strata contemporaneous, nor does the failure of that idenity necessarily indicate a different period of origin. When the fragments of rocks of one formation are included in those of another, we have evidence that the rocks to which these fragments belonged were formed, consolidated and fractured before the others were deposited. The conglomerates, or pudding-stones, are filled with worn, rounded fragments of other rocks; indeed all the sedimentary rocks consist of fragments, fine or coarse, of rocks older than themselves.

117. As the unstratified or *igneous* rocks occur in no regular order of succession, their *age* is not always easily determined. Their relations, however, to the stratified rocks furnish some intimations of their relative ages. When an igneous rock has passed through a stratum, causing dislocation or changes of structure, it is manifestly more recent than the stratum. A volcanic rock, as lava, may flow over strata producing its characteristic effects upon them, and subsequently other strata may be deposited upon it, accom-

modating themselves to its form, but experiencing no heating effects from it; we can, in such case, identify the period of its eruption, as the one which elapsed between the deposition of the two beds.

118. In A. D. 1680, Leibnitz divided all rocks into two classes—stratified and unstratified—in accordance with their origin. Subsequently Lehmann, a German mineralogist, classified the stratified rocks, as 1. *Primitive*—those which contain no animal or vegetable bodies; 2. *Secondary*—those which contained plants and animals; and 3. *Local*—those which occurred in limited localities. Werner

Fig. 67.

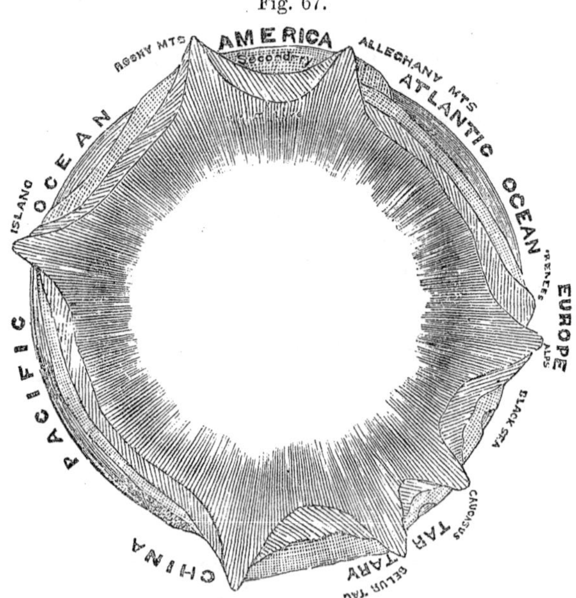

Section of the Globe.

made four classes: the *primitive*, *transition*, *flœtz* (flat-lying,) and *alluvial*. He proved that the stratified rocks overlie each other in a constant order of succession. He applied the term transition to the lower rocks which contain organic remains, indicating that the world was, during their deposition, passing from an uninhabited to an inhabited state; his *flœtz* rocks coincided generally with the *secondary* of other writers, and in the *alluvial* he included the most recent deposits. In the early part of this century, Cuvier and Brogniart proposed a new class called *tertiary*, and still more recently a fourth class called the *quaternary*, has been formed to embrace the diluvial or drift and alluvial deposits.

119. A section of the globe coinciding with the fortieth parallel of north latitude, shown in Fig. 67, exhibits the positions of the great mountain chains, with the stratified rocks sloping from and extended between them. The primary, transition, and secondary classes of rocks only are indicated, and the proportions of the globe and its "crust" are necessarily sacrificed; for those proportions refer to § 8.

120. A classification extensively used at the present day embraces all the stratified rocks in five classes.

I. In an ascending order, the first class is the *primary*, whose strata rest upon the unstratified, igneous rocks, are more or less crystalline in structure, and destitute of organic remains. This class includes the *metamorphic* rocks, described in § 108. Mr. Lyell designates these rocks, *hypogene*, nether-formed, because they have taken their present form at great depths.

II. The second class, the *transition* or *palæozoic*, are characterized by the remains of the earliest plants and animals. This class embraces the great coal formation.

III. The third class, the *secondary*, commences with

the new red sandstone above the coal, and extends to the top of the chalk.

IV. The rocks of the fourth class, the *tertiary*, are not generally so compact, nor so highly inclined, as the members of the first three classes. They contain the remains of many plants and animals, identical with living species.

V. The fifth class, the *quaternary*, includes the superficial deposits, the transported sand, gravel, clay, &c. of the drift, and alluvial, together with the local deposits of peat, marl, bog-ores, and soil formed by the disintegration of rocks in place, including the remains of animals identical with the present species, and some recently extinct.

121. The following tabular arrangement exhibits the five classes, together with their groups or systems of formations.

QUATERNARY,	Alluvium.
	Diluvium.
TERTIARY,	Pleistocene.
	Pleiocene.
	Meiocene.
	Eocene.
SECONDARY,	Chalk.
	Green Sand.
	Wealden.
	Lias.
	Triassic system.
TRANSITION or PALÆOZOIC,	Permian "
	Carboniferous system.
	Old Red Sandstone or **Devonian.**
	Silurian system.
	Cambrian "
PRIMARY,	Clay slate "
	Mica slate "
	Gneiss "

UNSTRATIFIED ROCKS—Granite.

122. The unstratified rocks do not admit of a systematic classification in accordance with a strict order of suc-

cession, but may be conveniently arranged in groups, depending upon the nature and mode of aggregation of their constituent minerals. Each of these groups is also associated with particular systems of the stratified rocks.

I. The *granite* group, comprising granite, syenite, serpentine, porphyry; of dense crystalline structure, and associated with the primary class, and the Cambrian and Silurian members of the transition class, of strata.

II. The *trap* group embraces basalt, green-stone, trachyte, amygdaloid; of a compact and less crystalline structure, and occurs in the upper transition or palœozoic, and secondary strata.

III. The *volcanic* group, less compact, vesicular, and associated with the tertiary, and quaternary deposits.

CHAPTER III.

PALÆONTOLOGY.

123. Most of the stratified rocks contain the remains of animal and vegetable bodies, which have been imbedded in them by natural causes. That branch of geology which classifies and describes these relics, is called *palæontology;* (*palaios,* ancient; *ontos,* being; *logos,* a discourse.) The bodies are called *fossils,* or *organic remains.* The discovery that particular fossils characterize certain deposits has greatly contributed to the rapid advancement of the science of geology. "Hence organic remains acquire a high degree of importance, not only from the intrinsic interest which they possess as objects of natural history, but also for the light they shed on the physical condition of our planet in the most remote ages; and for the invaluable data they afford as chronometers of the successive revolutions which the surface of the earth has undergone. They have been eloquently and appropriately termed the *medals of creation;* for as an accomplished numismatist, even when the inscription of an ancient and unknown coin is illegible, can from the half-obliterated characters, and from the style of art, determine with precision the people by whom, and the period when it was struck; in like manner the geologist can decipher these natural memorials, interpret the hieroglyphics with which they are inscribed, and from apparently the

most insignificant relics, trace the history of beings of whom no other records are extant, and restore anew those forms of organization which lived and died, and whose races were swept from the face of the earth, ere man, and the creatures which are his contemporaries, became its denizens."*

124. Fossils differ greatly in the *degrees of preservation* they exhibit. In a few rare instances animals have been preserved entire, with their flesh and skin. In 1774, the carcass of a rhinoceros was taken from the frozen sand of Siberia, with more hair on the skin than the rhinoceros of the present day has. At the commencement of the present century, the entire carcass of a mammoth was obtained

Fig. 68.

Siberian Mammoth found in frozen gravel.

from an ice-cliff in Siberia, twelve feet high, sixteen feet long, and with tusks nine feet in extent. The flesh was so well preserved that the wolves, bears, and hunters' dogs fed upon it. The skin was covered with a mixture of black bristles, fifteen inches long, and wool of a brown color.

---

* Mantell's Medals of Creation.

More than thirty pounds of the hair was collected. The brain and the capsule of the eye were in a good state of preservation. The skeleton, Fig. 68, together with a large quantity of the hair, is in the Museum of Natural History at St. Petersburgh. These animals are assigned to the pleistocene period, when they appear to have been numerous in that locality. Dr. Mantell states that thousands of fossil ivory tusks are annually collected there, forming a lucrative article of commerce, and that the remains of a greater number of elephants have been discovered in Siberia, than are supposed to exist at the present time all over the world.* Insects occur perfectly preserved, sealed up in amber, a fossil resin. Parts of the stomach and skin of large reptiles have in a few instances been found in older rocks, preserved by the antiseptic property of certain salts in the rocks. The "eatable earths" which the inhabitants of some countries have eaten mixed with saw-dust, consist of fossil infusoria. Usually, the harder parts of animals, the bones, shells, and crustaceous shields only, have been preserved.

125. In some cases no part of the animal or plant is preserved, but the space which the body occupied having been emptied by its decay is filled with mineral matter infiltered, and thus presents a perfect *cast;* or if mineral matter has not been infiltered we have only the *mould*. In a few instances impressions of only a part of the body, as foot-prints, are found. The tracks of birds occur in the New Red Sandstone above the coal, Fig. 69, though their skeletons have not been found below the chalk. Similar traces of other animals are met with in the same sandstone.

126. *Petrifaction* consists in the substitution of mineral

---

* Wonders of Geology, sec. II § 17

for organic matter. In some instances the animal or vegetable substance is almost entirely removed, while the organic structure is retained, so that thin sections examined with the microscope, show the forms of all the fibres and vessels

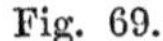
Fig. 69.

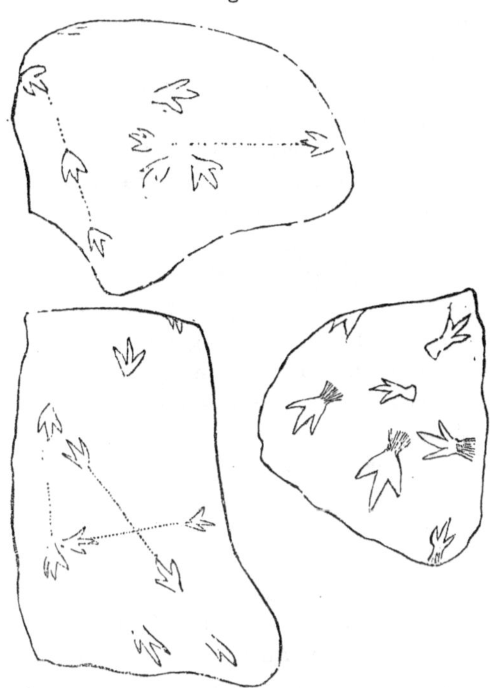

Bird-tracks in New Red Sandstone.

in their proper places. Limestone fossils placed in acids, have had all the petrifying material dissolved, and yet exhibited the animal tissues in a perfect form. The process of petrifaction has been imitated artificially; bones, leaves, &c.

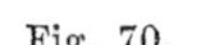
Fig. 70.

Tracks of the Cheirotherium.

have been buried in mud and sand, and after the lapse of a few years have been found petrified. The process is influenced by the presence of salts, as the sulphate of iron, in the mud, and is accelerated by heat and pressure. All fossils, however, are not petrified; nor does the age of the rock containing the fossil necessarily determine the amount of petrifaction: bones have been obtained from the Wealden, that were light and porous, while some from the most recent tertiary rocks were completely petrified.

127. Plants are sometimes petrified, but have more frequently sustained chemical changes, by which their own elements have been transposed and their vegetable structure destroyed; subjected to moisture and pressure, secluded from the air they ferment, evolve heat, and are converted into *bitumen*. This change is partially illustrated by a mass of half-dried hay, which ferments, becomes of a black color, and sometimes generates sufficient heat to take fire Bitumen is a black combustible substance, and liquid as petroleum, naptha; viscid as asphaltum, mineral pitch; or solid as jet, cannel and bituminous coals. In like manner animal muscle, buried in wet earth from which the air is excluded, is converted into a fatty wax called *adipocere*, (*adeps*, fat, and *cera*, wax,) retaining no trace of the original muscular fibre. In some of the fishes of the Old Red Sandstone, their muscles, blood, &c., have been converted into a dark-colored bitumen, which in some places pervades the rock to such a degree as to cause it to be mistaken for coal.

It resembles black wax, or when fluid, the coal tar of the gas works. This animal bitumen is eminently antiseptic, preserving in all their elasticity the bones, fins, and scales enveloped in it, better than the oils and gums applied to the old Egyptian mummies.*

128. The petrifaction of animal and vegetable bodies is frequently accomplished by means of the *metals*. The metallic salts, the sulphate of iron for example, dissolved in the waters of the earth, are decomposed; their oxygen uniting with some of the elements of the organic bodies, the metals are precipitated as sulphurets. Hence fishes are frequently found incrusted with iron-pyrites (sulphuret of iron) while their internal parts are converted into stone or bitumen; not unfrequently, however, their whole substance is changed into metal, all the traces of organic structure obliterated, and their form only preserved.

129. The most common *petrifiers*, are carbonate of lime, oxide carbonate or sulphate of iron, and silica. A fossil petrified in limestone, however, will not necessarily be calcareous. Many of the fossils of the chalk are flints, and those of clay-slates, calcareous. The cavities, as the interior of shells and hollow-bones, are often filled with crystals of limestone, or of silica, which, dissolved in water, was infiltered into these closed cavities through the pores of the shell or bone.

130. The *means* requisite for determining the characters of fossils, are furnished by such a degree of knowledge of *botany*, *zoology*, and *comparative anatomy*, as is adequate to the determination of living species. The same modes of investigation apply to both fossil and living species; they

* Miller's Foot-prints of the Creator.

are the natural complements of each other. Systems of botany and zoology are not complete without these extinct forms. In consequence of the relations which exist between the parts of animal frames, the observer is enabled to determine from a single bone the form and position of the other bones, and the entire condition of the animal. Thus the sharp, retractile, curved claws of the lion require that the bones immediately above them on the foot, and with which they articulate, be of such a shape as will allow free motion; the bones preceding these require also to be adapted to the design intended to be subserved by the claws. With these claws are associated the pointed sharp teeth adapted to tearing and cutting flesh. In animals feeding exclusively on flesh we always find the intestines about one fifth as long as in herbivorous animals. Hence a single tooth, or claw, suggests to the comparative anatomist the general form and habits of the animal. A single tooth of the Iguanodon enabled Cuvier to decide that the animal to which it belonged was an herbivorous reptile. The scales of fishes are so highly characteristic that Professor Agassiz has made their peculiarities the basis of his classification of these animals. A single scale found in the intestines of an Ichthyosaurus enabled him to identify the extinct species to which it belonged. In this way, peculiarities of structure reveal to the palæontologist the characters and modes of existence of creatures that ceased to exist ages before the creation of man, and to the geologist the condition of the world at that period.

131. The relics of animal and vegetable organization occur in almost every stratified rock, but in much greater *numbers*, and in a better state of preservation in some strata than in others. Nor are they equally abundant throughout

the same bed, some portions consisting almost exclusively of them, while in others they are sparingly diffused or entirely absent. Some limestones many feet in thickness are made up of shells or corals in countless numbers, cemented together by the carbonate of lime. But the greatest numbers of individuals accumulated in a limited space are presented by the most minute animals. Soldani obtained from a fragment of rock in Tuscany, which weighed one ounce and a half, ten thousand four hundred and fifty-four marine shells, resembling the Nautilus. One thousand of them weighed one grain. Some of the Wealden strata in England one thousand feet thick, and deposits in Auvergne seven hundred feet thick, covering an area eighty miles long and twenty wide, are filled with microscopic crustaceous animals. In Germany a bed fourteen feet thick, is made up of fossil animalcules so minute that, according to Ehrenberg, forty-one thousand millions of them occupy but a cubic inch. The total number of organic relics in the rocks is inconceivably great since more than two-thirds of the earth's surface is covered by fossiliferous deposits, many of which are several thousand feet thick. They are found at an elevation of at least seventeen thousand feet above the level of the sea, in strata of the Himalayas and Andes; and are obtained in excavations more than two thousand feet below the same level.

132. The medium through which fossils have been deposited has usually been the ocean; and by far the largest number are those which lived in the sea, and hence are designated *marine*. Some, however, lived in brackish water, and the strata in which they are found are termed *estuary deposits*, as was the case in the coal measures, and the Wealden strata. A part of the Tertiary series contains

fresh water animals and plants. Comparatively few *terrestrial* relics occur: but the Wealden, Tertiary, and Diluvial deposits contain some remains of *land* plants and animals.

133. Fossils usually occur in the situations in which they were at the time of their death, and hence often have their most delicate parts preserved. Some of the shells of the Wealden retain their epidermis, and ligaments of their valves, showing that they have not been subjected to friction. But there are immense reefs of broken shells and corals, some of which are worn to fine dust, indicating the violence to which they have been subjected.

134. In order to a full appreciation of the import of fossils, we must be enabled to compare them with each other at different geological periods, and with the existing races. This can be effected only by means of the distinctions and principles of classification used in Natural History.

A *species* is a group of individuals that are alike in every character not affected by accidental circumstances. Species are continued from generation to generation without the slightest change of character. Those variable changes which are produced by accidental causes, constitute *varieties*. Species are not convertible into each other. We know of no power adequate to the production of a species, short of the direct act of the Creator. Species have been obliterated in thousands of instances in the former ages of the world, and in a few known cases since the creation of man. (§ 86.)

A *genus* is an assemblage of species having certain characters in common. Genera usually embrace a considerable number of species, but a single species differing widely from all others may constitute a genus.

An *order* is an assemblage of genera, as a *class* is of

orders, while the whole are included in a *kingdom*. The terms *family* and *tribe* are used to designate groups intermediate between genera and orders.

135. We have seen (§ 87) that the plants and animals of the present period are distributed in floras and faunas, limited by climate and other circumstances. The examination of fossils in their beds shows that similar laws of *distribution* prevailed also at former periods. Classes and orders were, as they now are, very widely diffused, and genera often covered extensive areas, but species were restricted within comparatively narrow limits. In the periods of deposition of the earliest fossiliferous rocks, the temperature of the globe appears to have been higher than at present, and the faunas and floras less numerous, but covering a wider range. A formation existing in different parts of the world, very rarely contains the same species throughout its extent. The fossils of the chalk in England are in very few instances specifically identical with the American chalk fossils : the genera are the same, and the species analagous but not identical.

136. Examination of the successive formations of the fossiliferous rocks, shows that species of plants and animals have been introduced at different periods, continued for a time, then obliterated and their places supplied by others. The periods of duration of families, genera and species, are very variable. Species rarely extend through more than a single formation, but genera are often found in several successive formations, and in a few instances have survived all the mutations exhibited by the fossiliferous strata, having representatives in the existing races. Genera, however, and even orders are sometimes limited to a single formation ; in other instances they reappear after having ceased

through several formations. Those genera and species which survive through several formations, are such as have a wide geographical distribution, and seem to be endowed with a hardihood which enables them to endure changes of climate and other circumstances. Those species which live in deep water, with a more equable temperature are also more permanent. Shellfish and crustaceous animals are more enduring than vertebrate animals: no species of fish has yet been found in two successive formations.

137. Geologists presume plants and animals to have been introduced at the *period* of the deposition of the lowest strata in which their remains are found, though it is admitted they may have lived at an earlier period, and all trace of the existence of those first created, may have ceased in consequence of their frailty or the action of heat upon the lowest strata. Negative evidence on this subject is by no means conclusive. The failure to find the remains of a particular animal in a formation, does not satisfactorily prove that the animal was not living during that period: continued search may yet discover it; or its mode of life may have been such as to render the preservation of its remains after death, very improbable. But if from the remains found in a formation, we can infer the adaptation of circumstances to a certain class of animals, we do not expect to find in that formation animals whose organization require other circumstances. Thus in the coal formation whose plants clearly indicate the prevalence of a climate more elevated than that of the tropics at the present time, we do not expect to find indications of the existence of plants adapted to a temperate or arctic climate.

138. The close of geological periods appears to have been the *time* at which most of the *changes* in species oc-

curred; in some instances the change—obliteration of the old species and introduction of the new—is sudden, abrupt, and in others gradual. The introduced species are in no sense modifications of, or descendants from the species that preceded them. The facts of Geology disprove the hypothesis which supposes that animated existences were all introduced as mere points of vitality—monads—from which by a natural process of development, without the interposition of creative agency, they advance through continuous, regular grades, to the highest state of perfection and complexity. Mr. Hugh Miller has shown, on the contrary, that in certain orders there has been a degradation of species.* There was a time when reptiles represented the carnivorous and herbivorous quadrupeds, but such magnificent reptiles have not existed since. There was a time when birds seem to have been the sole representatives of the warm-blooded animals, but their foot-prints in the sandstone show, that in size no birds of the present day can compare with them. Although species remain constant in their characters, and changes are effected only by creative energy introducing new forms, still progress is discernible in the beings which have successively peopled the surface of the earth—progress in their approach to the present races. In the vertebrate animals the order of succession has been, fishes, reptiles, birds, mammiferous quadrupeds, and finally man. The relation between animals and the periods at which they live, appears to be a part of the plan of the Creator, as truly as is their distribution.

139. By a *comparison* of the fossils of different formations, we learn that those which are most unlike are most

---

* Foot-Prints of the Creator.

remote from each other in time. The organic remains of the Silurian and Carboniferous periods are more like to each other than they are to those of the chalk or tertiary; and the deeper we descend in the strata, the more unlike to the existing races are the fossils we obtain. In the earliest fossiliferous rocks the simplest forms of organization predominated, but representatives of all the classes are found with them. The relative number of species and genera of these classes has continually varied in the successive formations.

FOSSIL BOTANY.

140. All vegetables, fossil or recent, may be arranged under two grand divisions, *cellular* and *vascular*.

I. *Cellular* plants present the simplest forms of vegetation, consisting of an assemblage of cells of the same kind, without regular vessels, and having no visible organs of fructification. Such are the sea-weeds (confervæ, algæ,) mosses and lichens.

II. In *vascular* plants the cells are elongated into tubes or vessels, called *vascular tissue*, and form organs of nutrition and fructification. Four classes are formed in this division, in accordance with the structure of the organs of fructification.

1st. *Cryptogamiæ*, having neither perfect flowers nor visible seed vessels; to this belong the fern, equisetum or marestail, and calamites. Great numbers of these occur fossil.

2nd. *Gymnospermous Phanerogamiæ*, having flowers; but their principal characteristic is the nakedness of their seeds. To this class belong the Cycadeæ, or pine apple tribe, and the Coniferæ, or firs.

3d. *Monocotyledonous Phanerogamiæ,* flowering plants whose seeds have only one cotyledon, or seed-lobe, as the lily, the palm, and the cane. These are also called *Endogenous,* because they receive their growth wholly within, by the formation of new bundles of vessels at the center, crowding out and condensing those which lie toward the edge. They have no pith, concentric circles of woody fibre, nor true bark.

4th. *Dicotyledonous Phanerogamiæ,* flowering plants whose seeds have two cotyledons or seed-lobes, as the bean. These are *exogenous,* receiving annual deposition of woody fibre, upon the previous growth, and have true bark and pith. A cross section of the oak exhibits the pith at the center, concentric circles of woody fibre, connected by fine straight lines—the medullary rays—which radiate from the central pith. To this class belong our forest trees, and shrubs.

141. As fragments of trunks and branches are often the only specimens we have of a fossil species, we are obliged to resort to the study of their *anatomical structure.* These

Fig. 71.

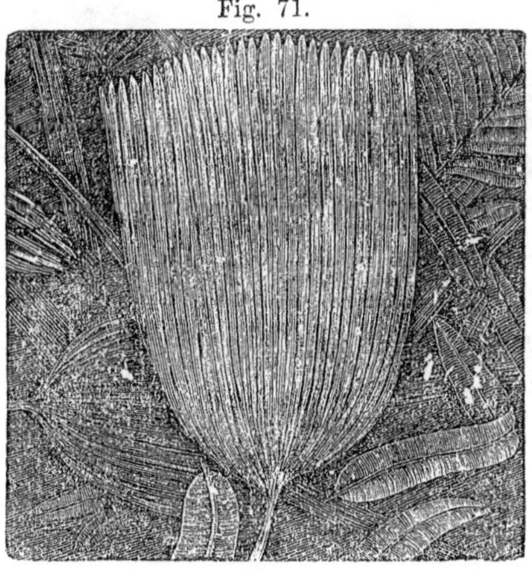

are quite distinctive and are frequently very perfectly preserved. A fragment of fossil wood cut very thin, will, with the aid of a microscope, show the form and position of the vessels so clearly as to enable the observer to determine the nature of the plant.

The *leaves,* either entire, or impressions of them are very common, and enable us to distinguish between endogenous and exogenous structures. If the veins of the leaf be all parallel, or connected only by little transverse bars, the plant was endogenous; but if they are unequal in thickness or arranged in net-like meshes, they belonged to an exogenous plant. The *flowers* of some liliaceous plants have been found with their corollas, and calyxes preserved, but the anthers and pistils could not be recognized. The preceding figure, 71, presents a portion of a coal plant, discovered at Tallmadge, Ohio, by Charles Whittlesey, Esq., which very closely resembles the corollas of some plants. The *fruits*

Fig. 72.

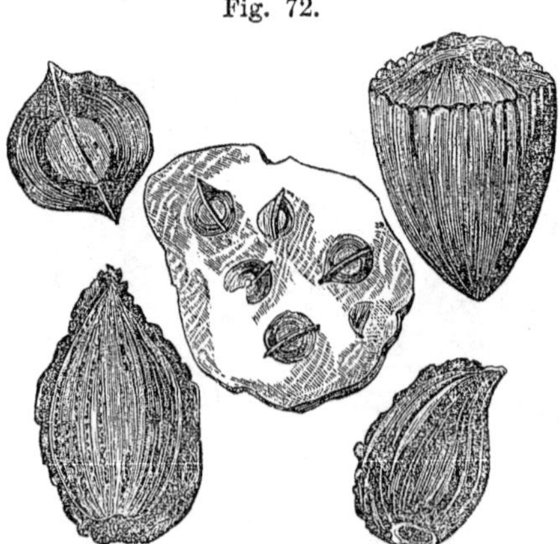

Fossil Nuts.

of many plants occur fossil in great profusion, as cones, nuts and seeds.

Some *trunks* of trees have been found of great length; as the Norfolk Island pine, forty feet long, in the carboniferous limestone of the Craigleith quarry near Edinburgh, and the "petrified forest" near Cairo in the Egyptian desert, where the silicified trunks of trees from forty to sixty feet in length and three feet in diameter, cover the ground. The resinous *secretions* of pines and other coniferous plants, are sometimes found fossil, as amber, which has been found in its natural position in trunks of trees. Some specimens of amber are black, showing the process of bituminization they have undergone. This also is supposed to be the origin of the diamond.

142. The cellular plants, as the algæ, occur most abundantly in the most ancient fossiliferous strata; in some of the Silurian rocks, entire layers are formed of a large species of sea-weed. Of vascular plants, the cryptogamous—fern, equisetum, and calamite—are found predominating in the carboniferous rocks, but existed earlier than that period. As we ascend in the series of rocks from the secondary class to the alluvial, the fossil ferns diminish in number, as they do when we travel from the equator toward the poles at the present day, when of the one thousand five hundred species, one thousand two hundred are within the tropics.

The gymnospermous plants—Cycadeæ and Coniferous—were most abundant in the lias, oolite, and tertiary periods. The endogenous class including the palms flourished during the tertiary period, though they are found to some extent in earlier periods. In the Illinois coal field a group of fossil palm trees has been discovered with their roots in the clay, and their trunks in the coal and sandstone above. The *geological position* of the exogenous plants is in the tertiary and recent deposits. The lignite or brown coal is almost entirely composed of dicotyledonous trees, the poplar, willow, elm, chestnut, and maple.

143. We have reason to presume that the history of fossil vegetation is not complete, since probably many families were too frail to be preserved, and many of the dicotyledonous class were not in situations favorable to their fossilization. The grasses, so abundant at the present time, seem to have been almost entirely absent from the secondary period, when plants allied to them were flourishing. The exuberant growth of vegetation during the coal period has led M. Brogniart to conjecture that carbonic acid was more abundant in the atmosphere at that time than at other periods. The number of fossil species of plants at present recognised and described, is about eight hundred.

144. The various *forms* which the carbonaceous matter accumulated by plants presents—anthracite coal, bituminous coal, jet, lignite or brown coal, and peat—depend upon the original structure and composition of the plants; upon the nature of the mineral deposits in which they were imbedded; and upon their more or less complete seclusion from the air. The conversion of wood and peat into lignite and coal, is still occurring. In the state of Maine, true bituminous coal has been found at the depth of four feet from the the surface, in a peat-bog of twenty feet thickness. Sections of the coal show that it is the remains of coniferous trees, immersed in the peat. The timbers of the Royal George, an English ship of war sunk off Portsmouth, and raised after sixty years immersion in the silt of the ocean, resembled in texture and color, the wood of submerged forests.

FOSSIL ZOOLOGY.

The number of animals fossil, as well as recent, is so great and their structure so diversified, that an outline of the classification of the animal kingdom is indispensable to a due appreciation of the Divine plan manifested in their creation and distribution.

145. According to Cuvier's Classification, the animal kingdom consists of four great divisions—the *Radiated*, *Molluscous*, *Articulated*, and *Vertebrated*.

146. I. The *Radiated* division is so called because the animals belonging to it have their organs of sense and motion arranged circularly, in radii from a center. They either have no nervous system, or one very indistinctly marked. Some are fixed as the corals (§ 74.); others move about as the star-fishes, and sea-urchins. The division of Radiata embraces three classes.

1st. Sea-urchins, whose surface is studded with spines—*Echinodermata*.

2nd. Jelly-fishes, including the Portuguese man of war and sun-fishes—*Acalephæ*.

3d. Polyps, attached to the earth like plants; many of them secrete coral—*Polypi*.

If sponges belong to the animal kingdom, they should be classed with the animals of this division. The crusts of the Echinodermata are well adapted to preservation and are found in many of the formations, and the secretions of the polyps—corals—abound in the rocks. The jelly-fishes, having no hard parts, have very rarely been preserved. Agassiz estimates the number of living species of Radiata at ten thousand. About one thousand five hundred extinct species have been recognised: they abound in the palœozoic—oldest fossiliferous—rocks.

147. II. The animals of the *Molluscous* division, have their nervous systems dispersed through their bodies in irregular masses; their muscles are attached to the skin, which in most of them is covered with a shell, and they have distinct systems of circulation and respiration. This division embraces three classes.

1st. Those whose arms—tentacles—are arranged about their mouths, as the cuttle-fish—*Cephalopoda.*

2nd. Those which creep upon a flat disc or foot, as the snails—*Gasteropoda.*

3d. Those which have no distinct heads, and are enclosed in bivalve shells, as the oyster and clam—*Acephala.*

As these animals lived in situations favorable to the preservation of their shells, we find great numbers of them fossil, and they are more frequently used to determine the identity, or relations of strata than any other class of fossils. Shells are either *univalve, bivalve,* or *multivalve,* according to the number of pieces which make up the shell. Some univalve shells are divided by tight partitions, and are called *chambered* shells. The branch of science which classifies and describes shells, is termed *Conchology.*

The Cephalopoda abound in the older strata; and the other two classes in the more recent rocks. The total number of living molluscous species is estimated at fifteen thousand. About six thousand extinct species are described.

148. III. In the animals of the *Articulated* division, the nervous system is arranged in the form of two parallel cords, which at intervals swell into knots, or ganglions. The bodies of these animals consist of joints—articuli—each one of which is furnished with a ganglion. This division embraces three classes.

1st. Insects, including spiders.

2nd. Worms, including leeches.

3d. Crustaceous animals, as the lobster and trilobite.

Insects are by far the most numerous class of animals, but as they are very perishable, and are devoured by many animals, few relics of them are found in the rocks: still they have been found as low in the series as the coal. Spiders are first seen in the oolite, and scorpions in the coal. Worms, with the exception of a few which secrete a calcareous covering, have left no trace of themselves in the strata. The entire number of species of the articulate is estimated at two hundred thousand. The number of extinct species now known is about eight hundred.

149. IV. The highest division of the Animal Kingdom is the *vertebrated*, whose members are characterized by a skull containing the brain, and a spine—vertebræ—containing the principal trunk of nerves. They have a skeleton, five senses, and red blood. Man belongs to this division. It embraces four classes.

1. Animals which suckle their young—Mammalia.
2. Birds.
3. Reptiles, including lizards, snakes, turtles, and frogs.
4. Fishes.

The fossil remains of animals of this division are very important, intimating most distinctly the circumstances in which they lived. The members of this division first introduced were the fishes, whose remains have been found in the lower Silurian and Cambrian rocks. Professor Agassiz distinguishes them, according to their coverings, into four orders. 1. Those with enamelled scales, like the gar-pike—*Ganoids*—(fig. 73.) 2. Those with the skin like shagreen, as the sharks and skates—*Placoids*—(fig. 74.) 3. Those which have the edge of the scales toothed, and

Fig. 73. Fig. 74. Fig. 75. Fig. 76.

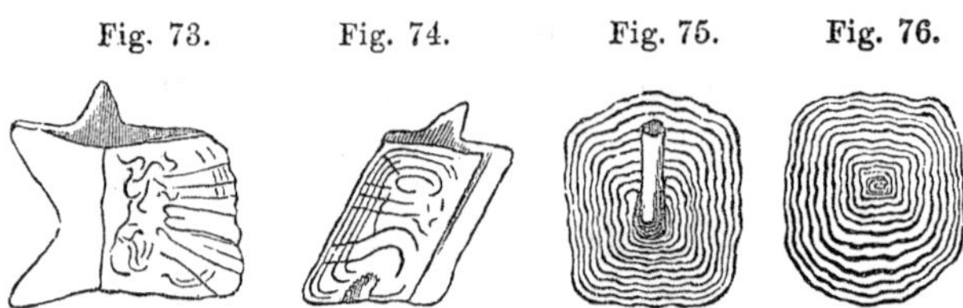

usually bony-rays in the fins, as the perch—*Ctenoids*—(fig. 75.) 4. Those whose scales are entire, and whose fin-rays are soft, like the salmon—*Cycloids*—(fig. 76.) The Placoids were first introduced; and the Ganoids succeeded; while Ctenoids and Cycloids were later, commencing at the close of the secondary period, and constituting three-fourths of the existing species.

The number of living species of fishes is eight thousand, and of extinct species now known, nine hundred.

Reptiles first appear in the New Red Sandstone, and become predominant in the Oolite. Their remains are among the most remarkable in the strata. Tortoises are found in all formations above the coal. The number of species of Reptiles is estimated at two thousand. Of the extinct species one hundred and twenty are known. The tracks of *birds* are found in the rocks abundantly, as low as the New Red Sandstone, and their skeletons in the Wealden, Tertiary, and Diluvian. There are supposed to be six thousand living species of birds, and thirty-five extinct species have been recognised.

The remains of *Mammalia* first appear in the Oolite, in which some species of Marsupials (animals furnished with a pouch for the protection of their young, as the Opossum) have been found, but the remains of animals of this class are found principally in the Tertiary strata There are probably two thousand living species of Mammalia; an two hundred and seventy-five extinct species are described.

150. The fecal remains of fishes, reptiles, and some other animals, are called *Coprolites.* They in some instances clearly indicate the nature of the food of the animal to which they belonged. The Coprolites of the Saurians contain the scales of fishes. The spiral convolutions of the intestines characteristic of the earliest fishes, are distinctly impressed on their fecal remains. Coprolites are found sometimes detached, and sometimes in the intestines of the fossil animals.

Fig. 77.

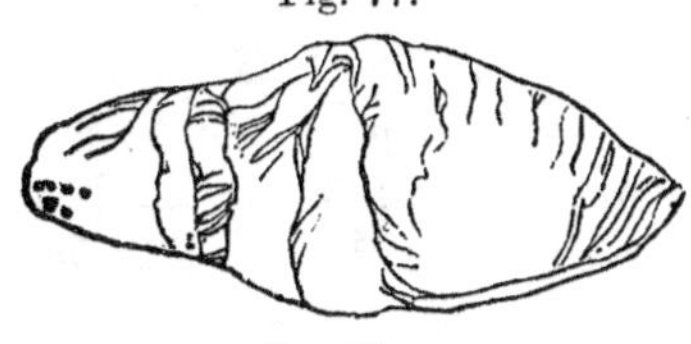

Coprolite.

151. The geological formations have sometimes been grouped and classified, in accordance with the predominant types of animals and vegetables which lived at the time of their depositions. The periods have accordingly been designated, as—*protozoic* (*protos,* first; and *zoe,* life;) the period in which are found the first forms of animated existence; *mesozoic* (*mesos,* middle; and *zoe,* life;) the period of intermediate forms of life; and *cainozoic* (*kainos,* new; and *zoe,* life;) the period of most recent living forms.

152. The division which Professor Agassiz has proposed, based upon certain classes of animals characteristic of the eras during which they flourished, is more definite.* He distinguishes four *Ages of Nature,* corresponding to great geological formations.

I. *The Primary* or *Palæozoic Age,* embracing the Cambrian, Silurian, and Old Red Sandstone systems: the

* Principles of Zoology, Chap. XIV. Section 2.

period during which no air-breathing animals lived; but Zoophytes, Shellfish, and Trilobites filled the seas. As fishes were the only representatives of the vertebrates, the age is called the *Reign of Fishes.*

II. *The Secondary Age,* comprising the Carboniferous, Trias, Oolite, (including the Lias,) and the Cretaceous formations. Insects, Reptiles, Birds, and Mammals, make their appearance during this period, but as reptiles preponderate, it is called the *Reign of Reptiles.*

III. *The Tertiary Age,* embracing the tertiary formations, in which the animals bear close resemblance to those of the present day, belonging generally to the same families, and very frequently to the same genera. Aquatic animals do not preponderate, as in the former ages, but large terrestrial mammalia abound; hence this age is designated the *Reign of Mammals.*

IV. *The Modern Age,* comprising all deposits since the Tertiary; characterized by the introduction of the most perfect animals, with man at their head. Some of these became extinct before the creation of man—as the mastodon. This age is denominated the *Reign of Man.*

CHAPTER IV.

THE UNSTRATIFIED ROCKS.

153. THE igneous rocks differ in chemical composition, in modes of aggregations and in position. They often pass into each other by insensible gradations. Two characteristic minerals, feldspar and hornblende, enter into the composition of all of them. When they have cooled under slight pressure they are porous in texture, as modern lavas; but when they have consolidated beneath great pressure, they are dense and crystalline, as granite.

GRANITIC ROCKS.

154. The *granite rocks* are highly crystalline, destitute of all traces of stratification, and occur as the basis upon which all the systems of strata repose; they are also thrust up to the surface in mountain masses, or injected into the strata in tortuous veins. They are associated, for the most part, with the oldest strata: they constitute the axes of principal mountain chains of the globe.

155. The characteristic ingredients of *granite* are quartz, feldspar, and mica, mingled in variable proportions. The prevalent color will depend upon the predominating ingredients; white and flesh-colored are common hues. The rock is either *coarse-grained*, when the distinct crystalline fragments of which it is composed are large; or *fine-grained*, when they are very small.

The quartz and feldspar are sometimes so arranged, as to present figures resembling Hebrew or Arabic letters. The rock is then called *graphic* granite. This structure is frequently exhibited by granite veins, which occur in older granite. When granite contains large distinct crystals of feldspar, it is denominated *porphyritic*. Granite appears, in some instances, as at Arran in Scotland, to have been protruded through the strata after it had consolidated.

Fig. 78.

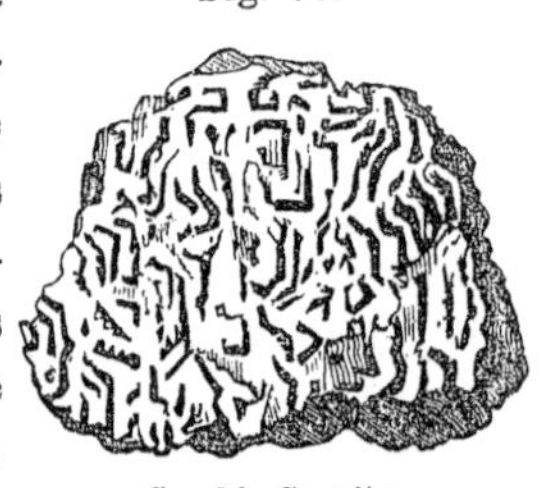

Graphic Granite.

156. In *Syenite* the three essential constituents are quartz, feldspar, and hornblende. It derives its name from Syene in Egypt, where it was formerly extensively quarried. The rock, composed of quartz, feldspar, mica, and hornblende, is called *Syenitic Granite*. Syenite is of a dark-green, or black color.

157. The term *porphyry* (*porphura*—purple) was applied by the ancients to a rock consisting of crystals of reddish feldspar imbedded in a base, of which feldspar was also an abundant ingredient. The rock was hard, susceptible of polish, and was used for archi tectural purposes, sarcophagi, &c. The term porphyry is now used to designate a rock of uniform compact base, through which are disseminated distinct crystals of any mineral. Crysals of fel lspar occur in basalt and trachyte, constituting

Fig. 79.

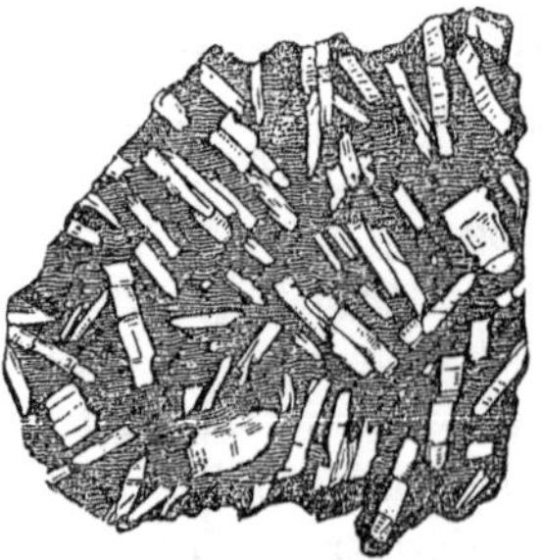

Porphyry.

basaltic porphyry, and trachytic porphyry. Instead of feldspar the imbedded crystals are sometimes of hornblende, or olivine.

158. *Metaliferous veins*, are most numerous in the primary and transition rocks; they rarely occur worth working above the coal. The ore does not usually occupy the whole vein, but is mixed with quartz, granite, porphyry, sulphate of baryta, &c., which constitutes the *vein-stone*, *gangue*, or *matrix*. Metallic veins differ greatly in thickness and extent; the contents of the same vein also vary in its different parts. Metallic veins are most productive when they pass through the junction of the unstratified and stratified rocks: their productiveness is also influenced by their direction; the east and west *courses* of the tin and copper veins of Cornwall are productive, while the *cross courses*—north and south veins—are not worth working.

TRAPPEAN ROCKS.

159. Certain crystalline rocks, composed of feldspar, and augite or hornblende, are called *trap rocks*, from the Swedish word *trappa*, a stair, because they frequently occur in tabular masses rising one above another, as terraces or steps. They appear to have been ejected as lava through fissures, and have produced the characteristic effects of heat upon the adjacent rocks. The trap rocks differ in constitution from the granitic, in containing much less silica, more magnesia, lime, alumina, and oxide of iron; and in being more fusible. The trap rocks occur in dikes, in dome-shaped masses covering other rocks, and in regular pillars; they are usually compact, but sometimes are full of pores. There are several varieties of trap rocks.

160. *Greenstone* consists of an intimate mixture of feld-

spar and hornblende of a dark green color, and is either compact or crystalline. When the rock is composed of albite and hornblende in grains it is called *diorite*. The structure of greenstone is sometimes massive, but more frequently tabular.

161. The constituents of *Basalt* are augite and feldspar When the former ingredient prevails, as is usually the case, the rock is of a dark green, or black color, which is due to the iron in the augite. Another metal, titanium, is usually an ingredient, and the mineral, olivine, is often found in grains or nodular masses in basalt. When coarse crystals of feldspar are disseminated in the rock, it is called *porphyritic* basalt. The structure of this rock is usually columnar.

162. Trap rock, containing almond-shaped cavities, is called *amygdaloid*, (from *amygdalum*, an almond.) These cavities are sometimes elongated by the flowing of the melted matter into cylinders several inches long. The cavities of the amygdaloid are filled with zeolites, quartz, calcareous spar, &c. A soft and earthy variety of trap, resembling indurated clay, is called *wacke* or *toadstone*.

163. *Trachyte* is composed of feldspar, with a small proportion of hornblende, titaniferous iron and mica. It derives its name from its roughness, or harshness to the touch—*trachus* in the Greek language signifying *rough*. It appears frequently in pulverulent masses of pumice, called *tufa* or *tuff* in rocks of all ages, invariably indicating the vicinity of erupted igneous rocks. Trachyte usually contains distinct crystals of feldspar, and consequently is porphyritic; it is also sometimes designated *trachytic porphyry*. A gray variety of basalt, consisting principally of feldspar, produces a clear ringing sound when struck with a hammer, and is called *clinkstone*.

164. *Serpentine*, a greenish rock containing a large proportion of magnesia, sometimes traverses the strata in dikes and veins, but occurs more frequently as a metamorphic rock.

165. Trap rocks very frequently present themselves in dikes, injected into fissures in the stratified and unstratified rocks. Fig. 64, §106, presents a powerful trap dike, penetrating the strata, in New South Wales. When such dikes pass through soft strata, as shale, they are sometimes left prominent, by the more rapid wearing away of the soft rocks. In some instances, however, they are decomposed more rapidly than the containing rock, in consequence of the oxidation of the iron in their composition; in such cases they leave open fissures.

166. The most conspicuous effects produced by trap rocks are the changes which their intense heat has effected on adjacent rocks. A trap-dike in Anglesea, passing vertically through shale, has altered the structure of the strata to the distance of thirty-five feet from itself, converting the shale into a porcellaneous jasper, and obliterating the fossil shells of the strata. In the north of Ireland basaltic dikes traversing the chalk have converted it into granular marble to the distance of ten feet. No traces of the organic remains, so abundant in the chalk, are discernible in the rock

Fig. 80.

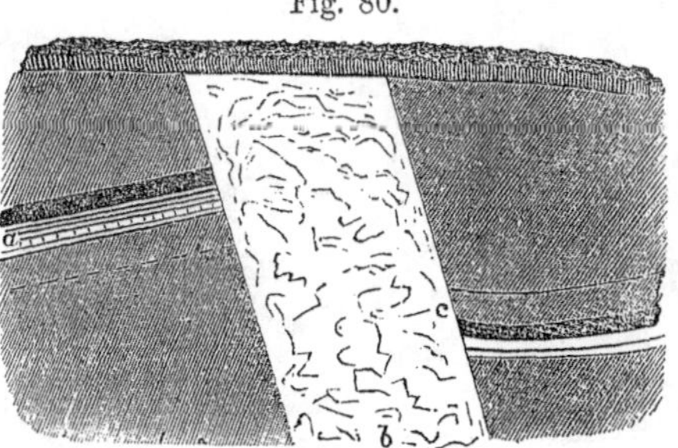

Trap-dike.

thus crystallized. Fig. 80 represents a trap dike displacing and altering the structure of strata. Secondary sandstones have been converted thus into quartz rock. Beds of coal, from their combustibility, exhibit the agency of these melted rocks in a striking manner. A greenstone dike passing through a coal-bed in Ireland has reduced it to cinders, through a space of nine feet on each side. In the north of England a similar effect has been produced by a trap-dike passing through the coal series. At a distance of one hundred and fifty feet, the structure of the coal is changed. At a certain distance it is coked, while in the immediate vicinity of the dike it is burned to soot. This dike extends seventy miles and is fifty feet wide.

167. Masses of trap rock are frequently met with, intruded between strata, covering extensive areas, closely resembling beds. They take this position when the resistance offered to their progress laterally between the strata, is less than that which they encounter from above. In some instances, having spread out in horizontal sheets upon the

Fig. 81.

The Giant's Causeway.

bottom of the ocean, they have been subsequently covered by sedimentary deposits.

168. The structure of the trap rocks, especially of basalt, is very frequently columnar, dividing immense masses into regular prisms, and constituting some of the most remarkable features of the earth's scenery. The "Giant's Causeway" is the legendary name applied to the most regular series of the magnificent basaltic columns which abound in the north of Ireland. The columns vary in the number of angles from three to twelve, but have most commonly from five to seven sides. They are divided transversely into joints, whose upper and lower surfaces are alternately concave and convex, as is shown in figure 65, § 107. These joints vary in length and diameter, but from six to eighteen inches are the prevalent dimensions. The surfaces of these

Fig. 82.

Basaltic Pillars of Staffa.

segments are so nicely adapted to each other, that neither between joints nor adjacent columns can a knife-blade be introduced; and yet each segment is completely dissevered from the others, so that it may be lifted out of its place. The Causeway extends under the sea. The neighboring headland seen at the left in Fig. 81, consists of alternations of basaltic columns, with beds of ochre interposed. As the ochre is worn away by atmospheric agencies, the columns are left unsupported, and many of them have fallen to the base of the precipice.

169. The Western Islands of Scotland consist almost exclusively of basaltic or trap rocks, which rise to the height of one hundred and fifty feet above the level of the ocean. The isle of Staffa is distinguished for its basaltic pillars and caves. The Scallop Cave, so called because the sides are formed of curved columns, extends one hundred and thirty feet into the rock, is thirty feet in height and twenty in breadth, at the entrance. But the principal

Fig. 83.

Fingal's Cave, Staffa.

object of attraction in this island is the Musical Cave, or as it is more commonly known by its legendary name, *Fingal's Cave.* This cavern is two hundred and twenty-five feet in length, with an average height of sixty-five feet; the breadth at the entrance is a little more than forty feet, and at its farthest recess is about twenty. The sides are columnar and the roof is partially so. The columns have been broken away in many places by the waves, presenting lateral recesses and jutting colonnades.

Fig. 84.

The Chimney—St. Helena.

170. Basaltic columns are not always vertical; they are sometimes horizontal, as is the case with "the Chimney" in St. Helena which is a portion of a trap dike. In broad flat masses of basalt the pillars are upright while in a vertical dike they are horizontal; and generally they are at right angles to the cooling surfaces.

Figure 85 represents curved columns radiating from each side of a bank of basaltic conglomerate. The basalt in contact with the bank exhibits no trace of columnar structure; but at a little distance regular columns appear, at first horizontal; after extending thus a few feet they curve gradually upward and at twenty feet distance stand nearly vertical.

Fig. 85.

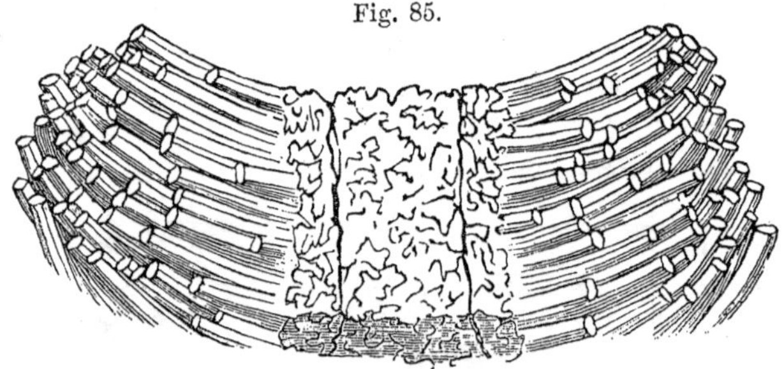

Basaltic Pillars—New South Wales.

The columns are from four to six sided, thirty feet long, and from one to four feet in diameter. The effects of heat upon the conglomerate are well marked. The melted basalt appears to have flowed over a small ridge of conglomerate in a stream fifty feet thick, and the forms and position of the columns were determined by the varying surface.*

171. The basaltic scenery of New South Wales delineated by Professor Dana, rivals that of the Scottish Isles. Some of the basaltic ridges rise to the height of four thousand feet above the level of the sea; and fissures or caves occur similar to those of Staffa. Figure 86 presents a view of the basaltic pillars at Kiama point. The sea enters through the channel seen in the cut, which is twenty feet broad and

* Prof. Dana, Geology of U. S. Exploring Expedition.

eighteen high; advancing by a subterranean passage for a distance of two hundred feet, it strikes against a wall of basaltic columns with deafening roar, and rises through an orifice to the height of one hundred feet. This is called a *blow-hole* or *spout-hole*.

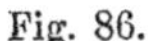

Fig. 86.

Kiama Blow-hole in Basaltic Columns.

172 Numerous examples of trap rocks and basaltic pillars occur in the United States—as the "Palisades" of the Hudson River, Titan's Piazza on Mount Holyoke, East and West Rocks near New Haven, Connecticut, &c. Vertical dikes with horizontal pillars in Máine and North Carolina present such regularity of figure and position as to have induced the belief that they were products of human skill. Trap rocks are associated with the metaliferous strata of Lake Superior and California, and successive rows of basaltic

columns, constituting mountain masses occur in the vicinity of the Columbia River, in Oregon.

173. The columnar and globular structure so characteristic of volcanic rocks, especially of basalt, early engaged the attention of Geologists and induced Mr. Gregory Watt to institute experiments in the year A. D. 1804 to ascertain its origin. He melted seven hundred pounds of basaltic rock, in a reverberatory furnace. He was enabled to effect its fusion with a less degree of heat than would have been required by an equal weight of pig-iron. When melted, it appeared as a dark, liquid, tenacious glass; a portion of it taken out and allowed to cool rapidly, retained its vitreous appearance, resembling obsidian. The remainder was left in the furnace, and occupied eight days in cooling; it was then cold externally, but still retained a considerable degree of internal heat. The interior of the mass, which had cooled slowly, exhibited a great number of spheres, which as they enlarged pressed laterally against each other, until they were cemented into regular prisms, with segments of alternate convex and concave surfaces. The articular structure and regular forms of basaltic columns seem therefore to have resulted from the crystalline arrangement of the particles while the rock slowly cooled under pressure.

174. The Feldspathic varieties of trap rocks, as trachyte, constitute the central or axial portions of mountains, while the exterior portions are composed of basalt; for instance, the summit of Mount Loa is clinkstone, while all its sides are basalt. This has been attributed to the difference in fusibility and specific gravity of the constituents, feldspar and augite. The temperature at which feldspar solidifies is sufficient to keep augite quite fluid. The augite also by

gradual cooling crystallizes and is converted into hornblende, for these two minerals differ only in crystallization, and that is determined by temperature and rate of cooling.

VOLCANIC ROCKS.

175. The term *lava* is applied to any rock which has flowed from a volcanic vent, but it is frequently restricted to the melted rocks erupted by active volcanoes or those recently extinct. The constituents of such lava are the same as those of the trap rocks, and we have *feldspathic* lava, and *augitic* lava, according to the predominance of either of those ingredients; but we do not find the compactness and crystalline structure so frequently exhibited, as in the older igneous rocks. Rapidity of cooling and difference of temperature and pressure, adequately account for the diversity of forms which igneous rocks present, while their constituents are essentially the same.

176. Lava bursting through the sides of volcanic cones presents itself in dikes similar to trap dikes; columnar structure also is exhibited by them.

When thoroughly melted lava cools rapidly, it resembles glass and is called volcanic glass, or *obsidian;* it is usually of a dark smoky color. Its fracture presents bright surfaces and sharp edges; it has been used by the Mexicans for mirrors, and knives. Lava of less glassy structure, and of a pitch-like luster is called *pitchstone.*

The cellular slaggy matter ejected from volcanoes is called *scoria;* when it is composed of feldspar, it is porous and fibrous and is then called *pumice.* It is so light as to float upon water. Scoria and pumice are the frothy scum upon the surface of melted lava, and become very abundant when hot lava comes in contact with water. Minute par-

Fig. 87.

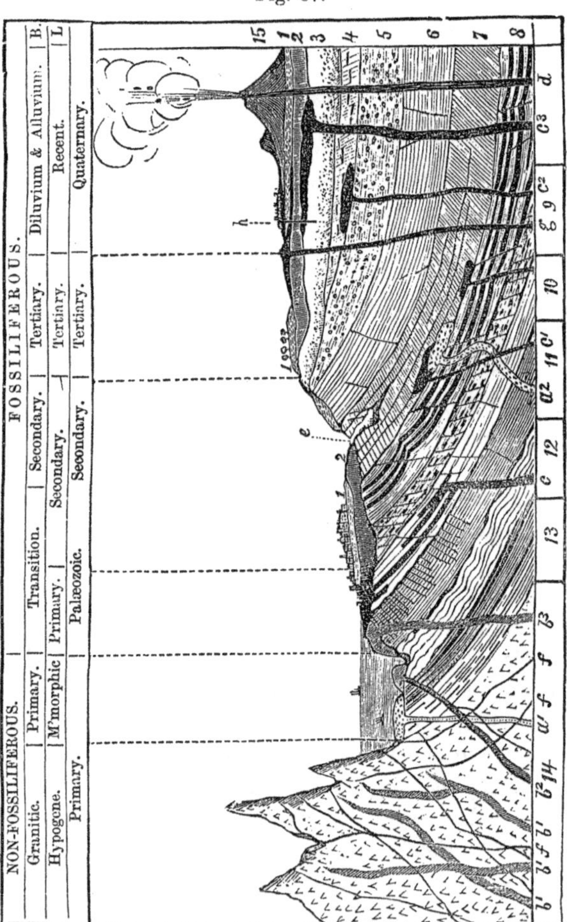

ticles of lava, light cinders, sand, &c., thrown out by volcanic eruptions, are called *volcanic ashes*. These materials cemented together constitute *peperino* and *pozzuolana*.

177. The figure on the preceding page presents an ideal section of the Earth's Crust, exhibiting the order of succession in the stratified rocks, and the relative positions of the unstratified.

B, signifies the classification adopted by Dr. Buckland; and L, that of Mr. Lyell.

1. Quaternary or modern deposits.
2. The deposits of the Tertiary.
3. The Chalk formation.
4. The Wealden.
5. The Oolite.
6. The Lias.
7. The Triassic system.
8. The Carboniferous system.
9. The Old Red Sandstone.
10. The Silurian.
11. The Cambrian.
12. Mica Slate.
13. Gneiss.
14. Granitic Rocks.
15. Volcanic Rocks.

a. 1. Porphyritic dike overlying the gneiss.
a. 2. Dike of Porphyry passing through the coal.
b. 1. Granite and Quartz Veins in Granite.
b. 2. Granite and Quartz Veins intersecting the porphyry, gneiss, and mica slate.
b. 3. Granite Vein passing into the transition rocks.
c. Greenstone dike of the transition period.
c. 1. Trap Dike intersecting a dike of porphyry.
c. 2. Trap Dike of the Oolitic period.
c. 3. Lava Dike of the volcanoes of the Tertiary period.
d. Lava of active volcanoes.
e. Cave in the magnesian limestone.
f. Metaliferous veins.
g. An extinct volcano.
h. An Artesian well.

CHAPTER V.

THE STRATIFIED PRIMARY ROCKS.

178. In describing the stratified rocks, we may begin with the most recent, in which the circumstances of deposition were very similar to those existing at the present time, and proceed to the most ancient strata whose origin was in a condition of nature widely different from the present, as regards both their mineral structure and fossil contents. But the chronological order appears to be more natural and satisfactory; commencing with the oldest strata, which are very widely extended, and more uniform in character, following the succession exhibited in nature, the facts may be delineated in the order of their occurrence. A knowledge of the phenomena of the older rocks is indispensable to a due appreciation of the characters of the more recent, since the latter are principally derived from the former.

179. The *primary stratified rocks*—the oldest sedimentary deposits—are composed of the same minerals that constitute the plutonic, granitic rocks, upon which they lie, and by the abrasion of which they were produced. As a class they are silicious in mineral character, and crystalline in structure; exhibiting, however, a stratified arrangement. Their particles appear to have undergone less attrition, and subjection to a higher degree of heat than the more re-

cent strata. No organic remains have been found in them; vegetables and animals either did not live upon the earth at the time of their deposition, or all traces of them have been obliterated by the heat to which the rocks have been subjected. Their thickness is variable, but usually very great; in some instances exceeding twenty miles. They are more widely extended than any other class of strata, and where they form the surface they present bold, rugged, and mountainous outlines. Their order of superposition is not invariable; gneiss, however, usually lies directly upon the granite, and mica-slate succeeds. This class of rocks has been denominated *metamorphic* (§ 109.) Veins and dikes of the igneous rocks, granite, porphyry, and greenstone, with metaliferous veins are frequently found traversing them.

Fig. 88.

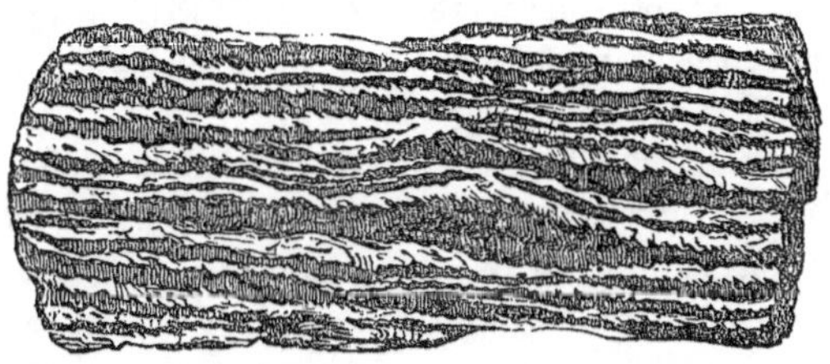

Gneiss.

180. The term *gneiss*—derived from a Saxon word signifying to sparkle—is applied to a crystalline aggregation of quartz, feldspar, and mica, distinctly stratified, in which the feldspar is less abundant than in granite, and the quartz is in fine grains. The mica gives to the rock a glistening aspect. Gneiss passes insensibly into granite, so that it is often difficult to say where the granite terminates and the

gneiss begins. The laminæ and strata are usually parallel to each other, and to the surface of the igneous rock upon which they lie, but frequently exhibit remarkable contortions, indicating a previous state of great flexibility. Dr. M'Culloch observes: "Imagination can scarcely conceive an intricacy of flexure of which a resemblance could not be found in the gneiss of the Western Isles of Scotland."

Distinct crystals of feldspar are sometimes found in gneiss; it is then termed *porphyritic*. When talc takes the place of mica, the rock is called *protogine*. If hornblende is added to the ordinary constituents of gneiss, the rock is termed *syenitic*. A section of graphic granite parallel to its imperfect laminæ, fig. 89, closely resembles a transverse section of gneiss.

Fig. 89.

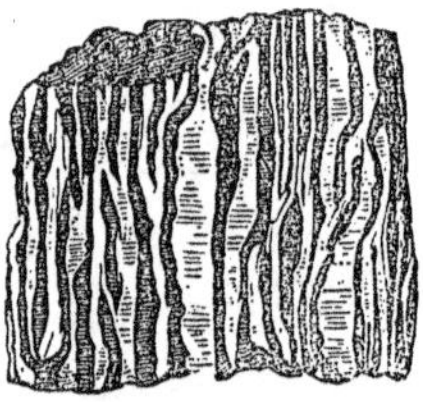

181. *Mica-slate* differs from gneiss, in the absence of feldspar, the greater predominance of mica, and the more perfect slaty structure; it often, however, passes gradually into gneiss. Perfect crystals of garnet and staurotide sometimes occur in this rock in as large proportions as either of its constituents, it is then denominated *garnetiferous* and *staurotidiferous*. Mica-slate passes by insensible gradations into clay-slate, and into talcose-slate.

182. *Hornblende-slate* is composed principally of hornblende, with a variable quantity of feldspar, quartz, and mica. When the hornblende and feldspar occur mixed in equal quantities the characters of the rock correspond with those of greenstone. Hornblende-slate occurs in all parts of the primary series, but is most frequently associated with mica-slate and clay-slate, into which it passes by imperceptible gradations.

183. The essential ingredient of *talcose-slate* is talc, which is sometimes pure, but more frequently mixed with feldspar, quartz, and mica. *Steatite* or *soap-stone* is a variety of this rock in which talc greatly predominates, is of a grayish green color, greasy feel, and is easily cut. *Chlorite-slate* is another variety, of a compact texture, and green color. Talcose-slate is a metaliferous rock, and frequently the repository of gold.

184. *Serpentine*, as stated in § 164, occurs both as a stratified and unstratified rock; in the former case exhibiting the characters of strata. As a metamorphic rock it is associated principally with talcose and hornblende slates.

185. *Primary limestone*, is a crystalline limestone, frequently pure white, and close grained, constituting the marble of sculpture; but is sometimes compact, and dark-colored. It occurs in thick beds, and in thin leaves. It alternates with gneiss, mica-slate, and clay-slate, and then usually contains some crystals of mica, quartz, and feldspar. It has been found in syenite, hornblende, and granite, entirely destitute of any trace of stratification, and even sometimes in the form of veins.

186. *Quartz rock* occurs interstratified with each one of the primary rocks. When pure it is granular, crystalline, with little or no tendency to divide in parallel beds; but when mixed with feldspar, mica, hornblende, talc, or clay slate, it exhibits regular stratification. In some instances it is eminently arenaceous—a mass of sand, exhibiting mechanical structure, and apparently made to cohere by the heat to which it has been subjected.

187. *Clay-slate*, or *argillaceous schist* is a rock common to the metamorphic and fossiliferous series. It consists of a very fine mixture of mica or talc with quartz, a large pro-

portion of argillaceous matter, a little iron, and some of the alkaline earths. Its prevalent colors are bluish-black, green, purple, and mottled, often accompanied by a glossy luster. Some varieties of slate are soft, and easily worn away, while others are hard and splintery. The slate which is immediately in contact with trap or other igneous rocks, has lost its fissile character having become flinty and changed its color. It often occurs interstratified with mica-slate and quartz rock, and in proportion to its nearness to granite or other igneous rocks, it is more highly glazed; it passes gradually into talcose, hornblende, or mica slates. Passing in the other direction, its luster becoming more dull and its texture less compact, it is called *shale*, and still farther up in the series it exhibits itself in beds of *clay*. *Roofing-slate* is a fine grained variety, of a bluish-black or purple color, splitting into thin, even slates. *Drawing slate* is a finer and more compact variety, containing some carbon. The *Whetstone* or *hone slate* is a fine-grained slate filled with minute grains of quartz, or grit.

188. Slates exhibit the phenomena of cleavage and joints with greater distinctness and regularity than any other rocks, and it is owing to this structure that they are

Fig. 90.

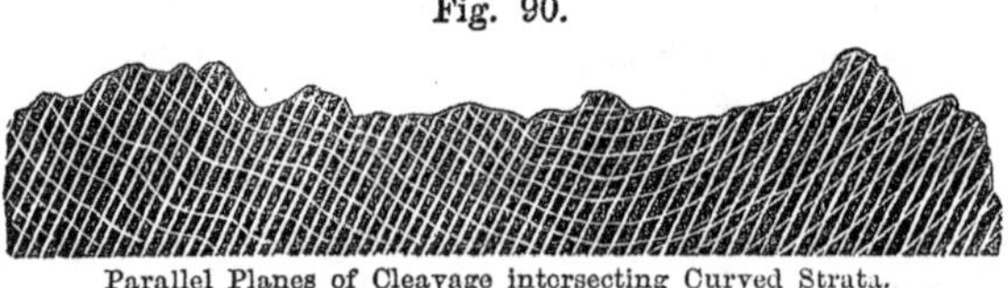

Parallel Planes of Cleavage intersecting Curved Strata.

capable of being split so evenly and extensively. Figure 90 presents the cleavage planes preserving their parallelism through contorted strata, and through alternations of fine-grained and coarse-grained slates, independent of the dip of the strata, (§111.)

Joints are distinguished from cleavage planes by the absence of all tendency in the rock to split between joints in a direction parallel to the joints, while the slate is capable of indefinite subdivision in the direction of the cleavage planes. In Figure 91, the flat surfaces, A B C, represent

Fig. 91.

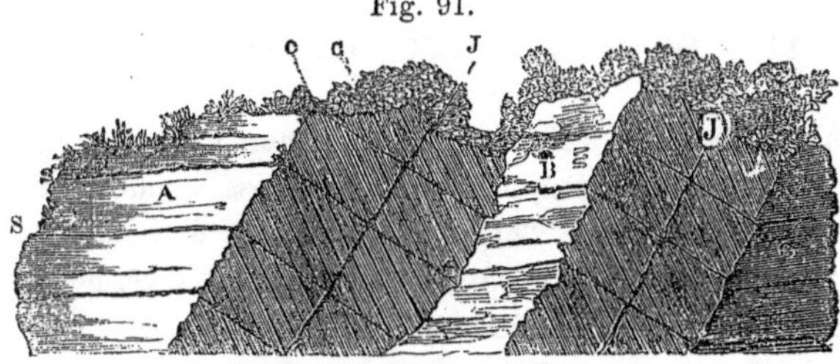

Stratification, Joints, and Cleavage.

exposed faces of joints, to which the walls of other joints, J J, are parallel: S S, are lines of stratification; C C, are lines of slaty cleavage intersecting the planes of stratification at a considerable angle,* (§110.)

189. The clay-slate is sometimes fossiliferous, especially as it occurs higher in the series. *Anthracite* occurs in primary limestone, mica slate and gneiss; and Mr. Lyell suggests that it is altered coal, since we observe coal in the vicinity of trap-dikes converted into anthracite. *Graphite, plumbago* or *black lead,* which is nearly pure carbon, is found in clay-slate and gneiss, and is supposed by some geologists to be coal still farther changed by the action of heat.

* Lyell's Elements.

CHAPTER VI.

TRANSITION OR PALÆOZOIC CLASS OF ROCKS.

190. THE rocks of this class are characterized by their wide geographical diffusion, and great thickness, amounting to many thousand feet. They were originally designated Transition by Werner, but are more appropriately termed palæozoic, since they contain the most ancient forms of living beings.

THE CAMBRIAN SYSTEM.

191. The transition from the crystalline and non-fossiliferous rocks, to the succeeding stratified fossiliferous, has been investigated in North Wales by Mr. Murchison and Prof. Sedgwick; hence the system is designated Cambrian, from Cambria or Wales. The rocks of this system are of enormous thickness, and eminently argillaceous in composition, varying from the finest clay-slate to coarse conglomerates. The term *grauwacke* (grey-wacke) used by German miners to indicate a grey rock, has sometimes been applied to the predominant member of this system, and to the system itself, inappropriately, since greywacke is a general term applied to rocks of different geological periods. The system is subdivided into three groups, taking names from the localities in which they are developed in Wales.

CAMBRIAN SYSTEM—THICKNESS ABOUT NINE THOUSAND FEET.

Plynlimmon Rocks—Slates, and conglomerates sěveral thousand feet thick.

Bala Limestone—A dark limestone, associated with slates, and containing remains of corals, shells, and fishes.

Snowdon Rocks—Fine grained, dark colored slates, and conglomerates, with remains of marine plants, fucoids, shells and zoophytes, several thousand feet in thickness.

The propriety of separating these groups from the following system—the Silurian—is questioned by some geologists.

THE SILURIAN SYSTEM.

192. The *Silurian system*—from the *Silures*, ancient Britons who inhabited Wales, where the rocks were first investigated and the system established by Mr. Murchison—embraces numerous rocks, consisting of sand, clay and limestone, abounding in organic remains, and giving clear indications of marine sedimentary origin. Some of the slates are very fine grained, while others are coarse conglomerates. The limestones are frequently highly crystalline.

UPPER SILURIAN—ABOUT FOUR THOUSAND FEET THICK.

The Tilestone—Hard, finely laminated, greenish sandstones, with reddish shales; organic remains abundant.

The Upper Ludlow Rock—Fine grained yellowish sandstones; beds of argillaceous sandstones filled with sea weeds, zoophytes, and scales, fins, jaws, teeth, and coprolites of fishes.

The Aymestry Limestone—A bluish gray limestone, mottled with white, containing numerous layers of shells and corals

The Lower Ludlow formation—Shales becoming slightly calcareous; sandy flagstones.

The Wenlock Limestone—Concretions of argillaceous limestones separated from each other by beds of shale; extremely fossiliferous.

The Wenlock Shale—Dark gray argillaceous beds, with calcareous concretions; remains of fishes.

LOWER SILURIAN—ABOUT THREE THOUSAND FIVE HUNDRED FEET THICK.

The Caradoc Sandstone—Finely laminated, greenish, fossiliferous sandstones; shelly limestones; remains of corals, mollusca, trilobites.

The Landeilo Flags—Hard, dark-colored, sandy flagstones, with beds of limestone; remains of trilobites, and mollusca abundant.

193. Although the subdivisions of the Silurian system established by Mr. Murchison, as existing in Wales, are not recognised in the rocks of that period, in other parts of the world, a similar arrangement is discernible in them, and his system is the universal standard of reference in comparing them. Silurian rocks, with essentially the same mineral characters and fossil contents, are found in Russia, Norway, Sweden, Asia, Africa and Australia. In North America, they are expanded on a scale of extent both lateral and vertical far superior to that of the Welsh rocks. One group of beds of shales and limestones, supposed to be contemporaneous with the Ludlow series of Wales, occupies a surface of more than 10,000 square miles in the valley of the Ohio. The Central and Western districts of the State of New York present these rocks on a magnificent scale, as detailed in the geological survey of that State. Specimens of these rocks have also been obtained within the Arctic Circle, and in Terra del Fuego.

194. Great interest attaches to the fossils of the oldest

palæozoic rocks, both because they are the most ancient forms of organized beings of which we have any knowledge, and are very unlike the present races. Representatives of all the classes of the animal kingdom are found in them. Of the vertebrata, fishes—Placoids and Ganoids—only have been discovered.

Trilobites are the principal representatives of the Crustacea, which are found in some localities heaped together in vast multitudes. These animals were of an oval figure, protected by a shield, covering the anterior parts of the body, while the abdominal portions consisted of segments, which enabled them to roll themselves into balls, so as to present their hard crusts in all directions in self-defense, as do the lobster and woodlouse of the present day. Their organs of locomotion, if any existed, must have been small, membranaceous or rudimentary, since no distinct traces of them have yet been satisfactorily made out. They were destitute of antennæ. The back presents two longitudinal furrows, dividing its surface into *three lobes;* hence the name Trilobite. Their eyes which are in many specimens perfectly preserved, giv-

Fig. 92.

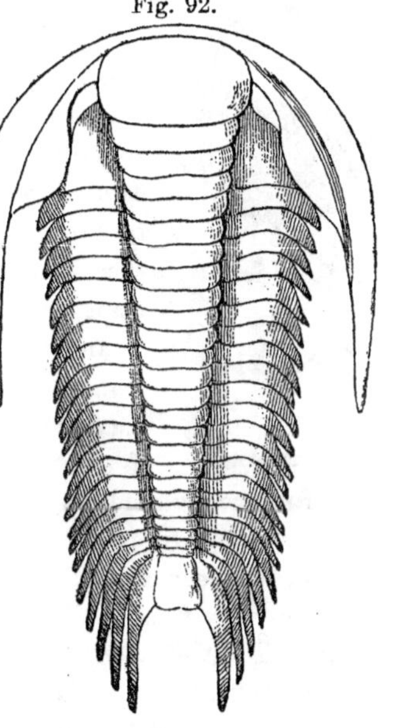

Trilobite.

ing us some indications of the condition of the early seas in which they lived, were compound, consisting of a great number of angular façettes or lenses, similar to the eye of the dragon fly, in which there are twenty-five thousand; and of the house fly, in which there are fourteen thousand façettes: the trilobite had several hundreds. In order to extend the field of vision, since the eye was immovable, it was elevated above their bodies. The form of the eye was that of a frustum of a cone, incomplete on the side opposite the corresponding part of the other eye, enabling the animal to see in all directions horizontally. This structure of the eye indicates that light bore the same relations to vision then, as it now does, and that these animals lived in water sufficiently transparent to transmit light. "We find in these animals," says Dr. Buckland, "an optical instrument of most curious construction, adapted to produce vision of a particular kind, created in the fulness of perfection, and fitted for the uses and condition of the class of creatures to which this kind of eye ever has been and still is appropriate." Trilobites vary from one to twenty inches in length. They were very abundant among the earliest inhabitants of the globe, and continued into the carboniferous period, after which no trace of them is discerned. Two hundred and thirty species, under several genera of the family Trilobite, have been

Fig. 93.

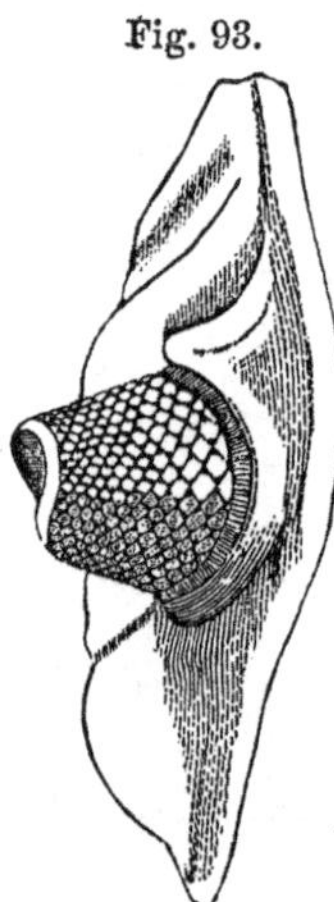

Eye of the Trilobite.

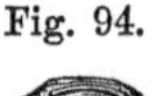

Fig. 94.

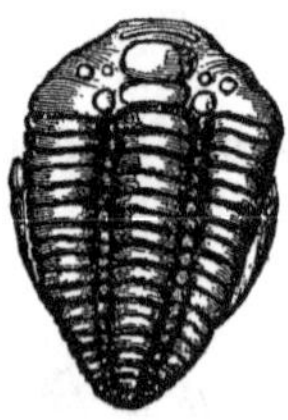

Calymene.

described; of which the genus Calymene, fig. 94, is found in the Wenlock division of the Upper Silurian rocks.

195. Of the Mollusca, the most numerous representatives in the Silurian rocks belonged to the class Cephalopoda (§ 147) including the *Ammonite* and *Nautilus*, and to the order Brachiopoda of the class Acephala. The latter inhabited bivalve shells, the valves of which, however, were not secured by a hinge, but were confined by a bundle of muscular fibres attached to one shell and passing through an aperture in the beak of the other; this gives to the interior of the shell a peculiar appearance. Several hundred species of the Brachiopoda have been discovered in these rocks. Among them was the genus *Terebratula;* created among the first, it has survived longer than any others, having several living species in the seas of the present day. The genera *Leptæna* (*Producta,*) *Delthyris, Orthis* (*Spirifer,*) and *Pentamerus,* belong to this period.

The *Orthoceratite* (*orthos,* straight; *keras,* horn,) was a straight shell, divided into numerous chambers by septa or partitions, which were perforated so that a tube (siphuncle) might communicate with all of them.

Fig. 95.

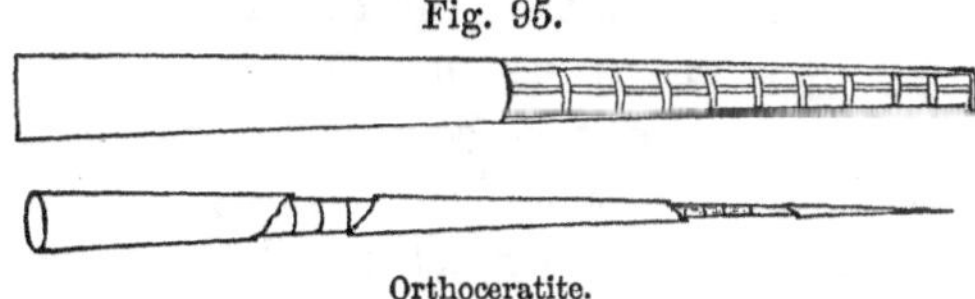

Orthoceratite.

196. Of the Radiata of the Silurian period, the coral polyps were the most numerous; their fossils form a very large proportion of the whole mass of the limestones, crowded together as are the corals of coral reefs in the existing oceans. The Chain-coral—*Catenipora*—is a beautiful ge-

nus characteristic of the Upper Silurian beds. In the lowest beds, and among the earliest forms of life, are found great numbers of a fossil, called, from their markings upon the rocks, *Graptolites*, supposed to have been zoophytes,

Fig. 96.

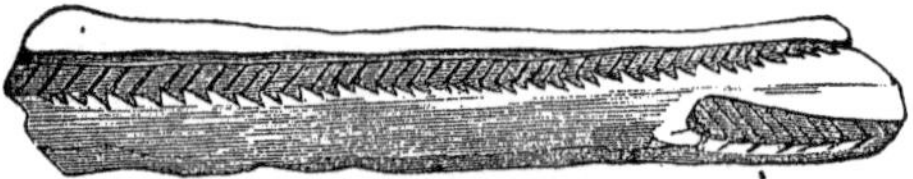

Graptolithus Ludensis.

allied to the sea-pens which inhabit the muddy sediment of the deep waters of the existing oceans. The Graptolites are not found in the limestones, but in the flags or slates of the Cambrian and Silurian rocks, and especially in a *mud-stone*, which indicates tranquil deposition. The *Crinoid* or *Encrinite* family is represented in this group of rocks, but was more fully developed at a later period. The fossils of the Silurian rocks indicate that they were deposited in deep seas, and in a temperature at least as high as that of the tropical regions of the present period. Marine plants were abundant, but the existence of land plants or animals at that period has not been established.

Fig. 97.

Graptolithus Murchisoni.

197. The igneous rocks associated with the Cambrian and Silurian systems, are granite, porphyry, and greenstone, whose incursions have often given to the strata inclined and contorted positions. But the Silurian beds of Sweden are horizontal, and those of the State of New York are but slightly inclined.

THE OLD RED SANDSTONE—DEVONIAN SYSTEM.

198. This system, consisting of conglomerates, sandstones, marls, and limestones, is intermediate in all respects between the Silurian and Carboniferous systems. The prevalent color of the marls and sandstones—dull red—is due to the peroxide of iron. It is called the *Old* Red Sandstone, to distinguish it from the *New*, which lies above the coal. It is extensively developed in Herefordshire and Devonshire in England, and in Scotland, where it has been fully investigated.* Although the beds of Devonshire differ from those of Herefordshire and Scotland, in mineral and mechanical condition, the characters of their fossils clearly show them to be contemporaneous with the Old Red Sandstone of the other districts; hence the system is sometimes called *Devonian.*

199. Tabular arrangement of the English and Scottish members of the series:

OLD RED SANDSTONE—ABOUT TEN THOUSAND FEET THICK.

ENGLAND.	SCOTLAND.	
Old Red Conglomerate	Quartzose Yellow Sandstone, Impure Limestone, Gritty Red Sandstone.	UPPER.
Cornstone.	Gray fissile Sandstone. . .	MIDDLE.
Cornstones and Marl,	Red and variegated Sandstones, Bituminous Schists, Coarse gritty Sandstone, Great Conglomerate.	LOWER.

In Scotland the base of the system is an extensive, thick

* "A New Walk in an Old Field—The Old Red Sandstone, by Hugh Miller, Esq.," is a work by a man of self-taught genius, combining the accuracy of scientific research with the fascinations of romance.

conglomerate; this is succeeded by coarse red and yellow sandstones alternating with green and red marls; upon this lie very thick beds of calcareous, micaceous and bituminous schists, loaded with fossil fishes, and containing impressions of plants.

The middle group, corresponding to the *Cornstone* of England, consists of a bluish gray sandstone and beds of friable stratified clay, while the contemporaneous English beds are principally red and green marls.

The upper part of the formation presents beds of sandstone alternating with limestone barren of fossils, and containing masses of exceedingly hard chert.

200. The subdivisions of the Devonian System, in Scotland and England are local, but serve as standards of comparison for the rocks of the same system in other parts of the world. They have been investigated as developed in Ireland, Belgium, and Russia; and are known to exist in South America and Australia. In North America, their essential characteristics, mineral and organic, have been shown to be identical with those of the Scottish series; they are developed in the States of Pennsylvania and New York, and west of the Alleghanies, in some places from one thousand to fifteen hundred feet thick.

201. The organic remains of this period are very peculiar, consisting of corals, molluscous and radiated animals, not differing widely from those of the Silurian system; but the vertebrate animals are here represented by a large number of species of singular fishes, in many instances so well preserved as to exhibit their outlines distinctly. These fossil fishes are sometimes denominated *ichthyolites*, (*ichthus* a fish; *lithos* a stone,) and although traces of them are discerned in the Silurian rocks, their abundance and perfect

preservation in the Devonian system have rendered them characteristic of it. They belonged to the Ganoid and Placoid orders of fishes, (§ 149.) Of the eighty Ganoid species hitherto described, more than fifty belong exclusively to the Old Red Sandstone formation. The Ganoid fishes are covered with angular scales of bone, coated with enamel, which regularly arranged invest the whole animal. Many of these fishes, in the structure of their teeth and other peculiarities approximate the reptiles, and hence are called *Sauroid.* The Ganoid species are nearly extinct; the sturgeon and bony pike are, however, living representatives of the order. Another feature of the fishes of this period approximating them to the Saurian type, is the prolongation of the vertebral column into the caudal fin, producing a *heterocercal* tail, as represented in figure 98. In all fishes of the existing seas, with the exception of the sturgeon, shark, and bony pike, the vertebral column terminates at the point, from which the caudal fin is given off equally above and below, constituting a *homocercal* tail, as exhibited in figure 99. All the fishes of the Palæozoic period had unequally bilobate or heterocercal tails.

Fig. 98.

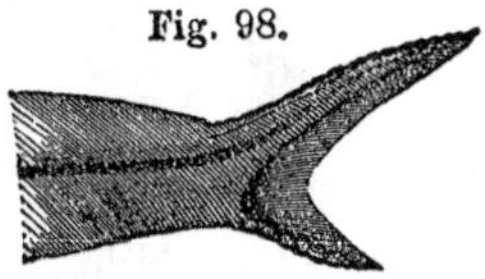

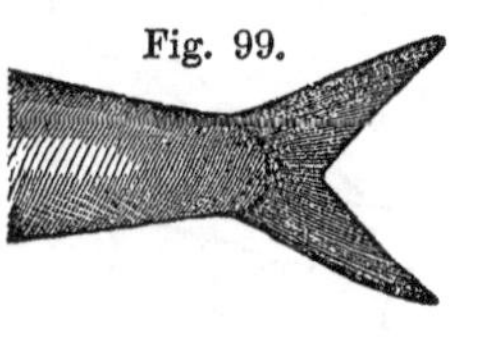
Fig. 99.

202. The *Cephalaspis* was a fish, whose most striking feature was an enormous shield forming the head; hence the name (*kephale,* head; *aspis,* shield.) This crescent-shaped buckler gave the animal the appearance of a trilobite, and the first discovered specimens were supposed to be crustaceous. The body was comparatively small, elon-

gated, terminating in a heterocercal tail. It was furnished with fins; the dorsal fins are usually conspicuous. The body and head were covered with bony, enamelled scales. This fish seems not to have attained a large size; specimens are about seven inches in length.

Fig. 100.

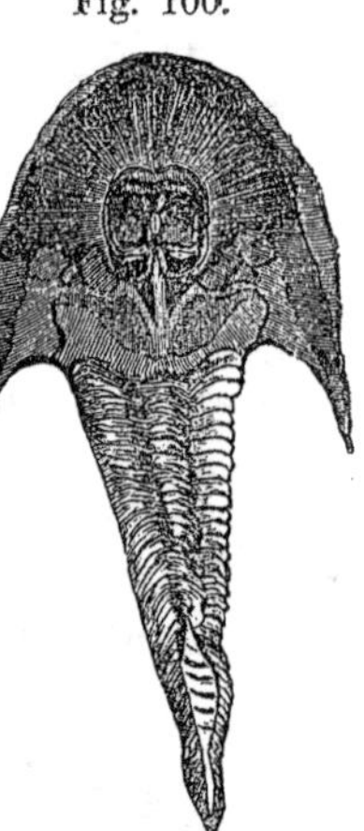

Cephalaspis Lyellii.

203. Another equally singular genus was the *Pterichthys*—the winged fish (*Pteron*, a wing; *ichthus*, a fish,) characterized by its two wing-like lateral appendages, which were its weapons of defense, and may have assisted in locomotion. Its body was protected by strong, bony, enamelled plates; its head was small, body flat, and its tail thickly covered with scales. It did not exceed a foot in length.

Fig. 101.

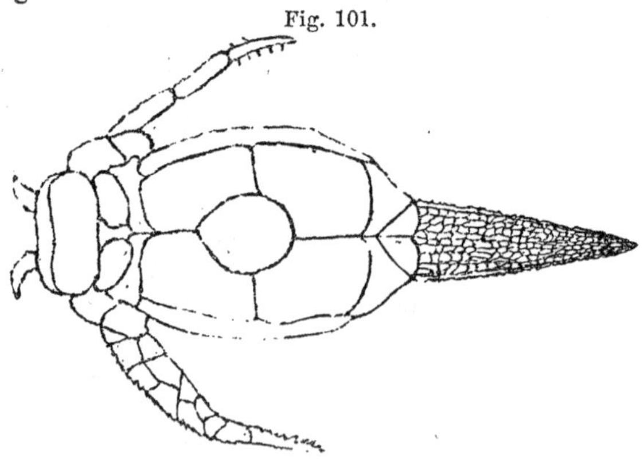

Pterichthys Cornutus.

Fig. 102.

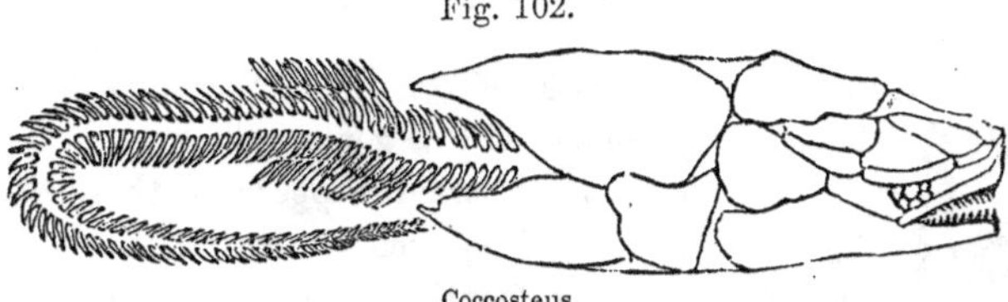

Coccosteus.

204. The *Coccosteus* resembles the Pterichthys in general structure and covering, but was destitute of wing-like appendages. It derives its name from the berry-like tubercles which stud its enamelled bony plates (*kokkos* a berry; *osteon*, a bone.) Its head was large, covered with several plates, and attached to the body by small articulating surfaces. The upper part of the body was covered with one plate and the lower part with four. The tail was much longer than the body and furnished with two fins. Many other genera belonging to this period have been described, among which may be named the *Holoptychius*, whose bony plates are more numerous and smaller; and the *Osteolepis*, in which the bony plates become scales and the figure of the animal approximates that of existing fishes. Some of these attained a large size.

205. The brachiopodous genera of shells, Terebratula, Orthis, Spirifer &c., except the Pentamerus, of the Silurian period are extended into that of the Old Red Sandstone. The corals *Cyathophyllum* and *Favosites* also are common to the two periods. The Favosites consisted of a congeries of diverging ascending tubes divided by lamellæ and communicating with each other by lateral openings. Figure 104 presents a polished section magnified to show the interior

Fig. 103.

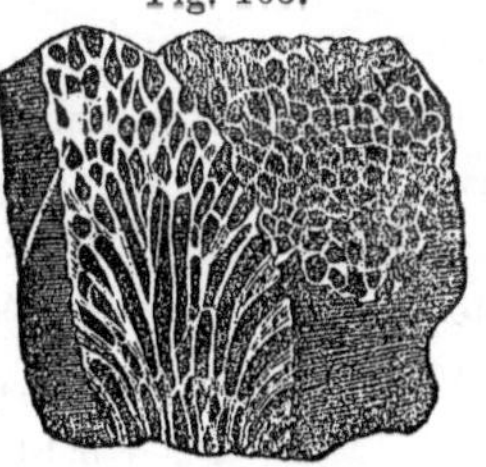

Favosites Polymorpha.

Fig. 104.

structure. The rocks of New York and Ohio abound in specimens of this genus. The trilobites of the previous period reappear in the Devonian, in several species of which the *Brontes* grew to the length of four feet, and closely resembled the lobster, having similar tuberculated claws.

206. The plants of the Devonian period appear to have been mostly marine fucoids, which are so numerous in some of the straty, that the slate beds owe their fissile character to the layers of carbonated sea weeds. Ferns and plants allied to the *Lepidodendra* of the carboniferous period occur; and Mr. Miller has recently proved that Dicotyledonous gymnospermous plants with true woody fibre existed at that time.*

207. The igneous rocks associated with these strata, are principally varieties of the trap, which have invaded and distorted many of the beds. At the close of the Devonian period, the true granitic eruptions appear to have ceased.

THE CARBONIFEROUS SYSTEM.

208. The Carboniferous system consists of a widely extended series of fossiliferous limestones, alternating with sandstones and dark bituminous shales, containing beds of coal. The subdivisions of the system in England are represented in the following table.

CARBONIFEROUS SYSTEM.

The COAL MEASURES, 3000 feet thick.	Sandstone, Shale, and Coal, alternating with beds of clay and ironstone. Land plants in profusion. Limestone, with fresh-water and marine shells.

* Footprints of the Creator, p. 222.

MILLSTONE GRIT, 600 feet thick.	Sandstone, Shale, and Conglomerate, Shale and thin seams of coal—coal plants.
CARBONIFEROUS, or MOUNTAIN LIMESTONE, 1800 feet thick.	Limestone and flagstone abounding in corals and marine shells; with layers and nodules of chert. Ores of copper, zinc, lead, barytes, and fluor spar. Limestone, with innumerable shells—*spirifer, goniatite, orthoceratite, bellerophon,* &c. Varieties of black, bluish gray, and variegated marbles.

This system of rocks is developed in Ireland, France, Belgium, Spain, Russia, China, and America.

209. The organic remains of this period are of marine, fresh-water, and terrestrial origin. *Ichthyolites*—teeth, fins,

Fig. 105.

Fossil Fish, Ohio.

spines, scales and coprolites of fishes, as well as the entire fish abound. The fishes belonged to the ganoid and placoid orders—sauroid—and in some instances, as in the Megalichthys, were of gigantic size.

Of the *crustacea,* the trilobites were still the principal representatives. Shell fish were very numerous and large; representatives of all the orders—multivalve and univalve, single-chambered and many chambered occur aggregated in masses, as they lived gregariously. Among the bivalves the Terebratula, Spirifer, and Inoceramus; and among univalves, the Goniatite, Orthoceratite, Bellerophon, Euomphalus and Ammonite, were frequently recurring genera.

Fig. 106.

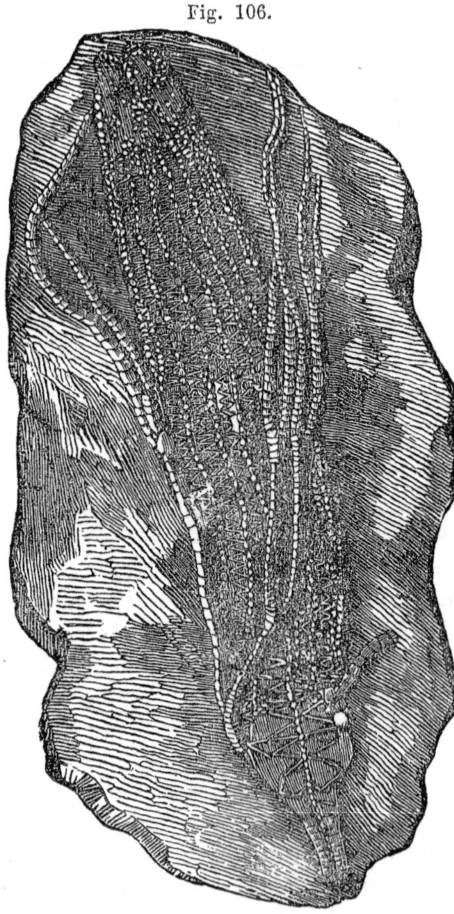

Pentacrinite.

A few specimens of Coleopterous insects—beetles, and of the scorpion have been found in this series. Of the *radiata*, the *crinoid* or *encrinite* family, of which the Pentacrinite is a member were most abundant; (§ 223) some of the limestones of this group are composed almost exclusively of their remains, and are designated *encrinal*. The coral polyps formed extensive reefs. The marine fossils of this period are mostly confined to the lowest member of the system—the carboniferous limestone, in which they are estimated to form at least three-fourths of the whole mass.

210. In the organic remains of the upper members of the system—the *coal-measures*—the great abundance of vegetable fossils, indicating a most luxuriant growth of land plants, is the most striking feature. That coal is of vegetable origin, is abundantly proved by the organic structure which can be seen in it with the aid of a microscope. The differences observed in coals may be due to their origin from different plants. The coal occurs in strata of sandstones and shales, with intervening layers of ironstone and limestone, some of which appear to be of fresh-water, and some of marine origin. The beds of coal vary in thickness from the fraction of an inch to thirty feet; in one of the English coal-fields their aggregate thickness is one hundred and fifty feet. The number of beds also varies, sometimes exceding one hundred; the central beds are usually the thickest and purest. The coal-measures are often designated as *basins* because they present themselves in troughs or basin-shaped cavities

211. The shales which underlie and overlie the beds of coal contain the best specimens of the coal plants, which occur between every succession of laminæ. The newly exposed roof of a coal mine presents a beautiful display of interlacing stems and leaves. "The most elaborate imitations of living foliage on the painted ceilings of Italian palaces bear no comparison with the beauteous profusion of extinct vegetable forms, with which the galleries of these instructive coal-mines are overhung. The roof is covered as with a canopy of gorgeous tapestry, enriched with festoons of most graceful foliage flung in wild irregular profusion over every portion of its surface. The effect is

heightened by the contrast of the coal black color of these vegetables, with the light groundwork of the rock to which they are attached. The spectator feels transported, as if by enchantment into the forests of another world; he beholds trees of form and character now unknown upon the surface of the earth, presented to his senses almost in the beauty and vigor of their primeval life; their scaly stems and bending branches, with their delicate foliage are all spread forth before him, little impaired by the lapse of indefinite ages, and bearing faithful records of extinct systems of vegetation, which began and terminated in times of which these relics are the infallible historians. Such are the grand natural herbaria, wherein these ancient remains of the vegetable kingdom are preserved in a state of integrity little short of their living perfection, under conditions of our planet which exist no more."*

212. Several hundred species of coal plants have been discovered which are allied in their generic characters to the ferns, canes, pines, cacti &c., but as species, became extinct before or at the close of the carboniferous period. The same species are mostly found in coal beds in Europe, Australia, the United States, and in Melville Island in 75° north latitude. The plants of the present period which most resemble the coal plants, live within the tropics, and are very much smaller than the corresponding members of the Flora of the carboniferous era. One of the most common fossils of the coal strata is the *calamite,* so called from its reed-like appearance. It was cylindrical, gradually tapering to a point, and frequently curved at the extremity; ribbed or furrowed longitudinally, and surrounded at intervals by horizontal rings or articulations. It was a branching plant with a hollow stem, woody tissue, and distinct bark. It has

* Dr. Buckland's Bridgewater Treatise, p. 458.

Fig. 107.

Tree Fern.

been supposed analogous to the equisetum or horse-tail of the present day, but was much larger, attaining the size of fourteen inches diameter.

Fig. 108.

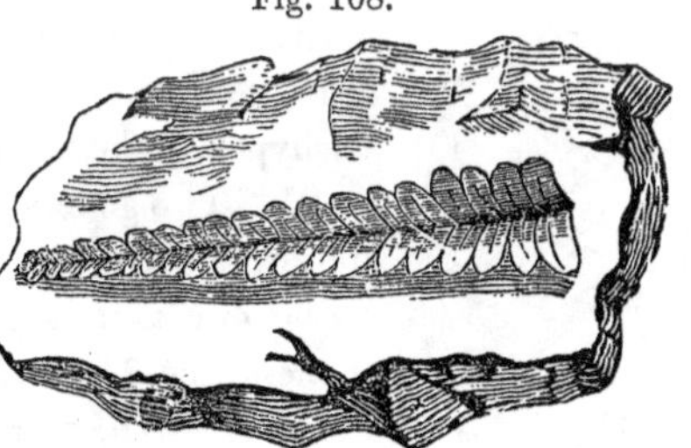

Lonchopteris.

213. The fronds or leaves of the *tree-fern* preponderate over all other fossils in the shales and sandstones of the coal measures. Their trunks are marked with scars, cicatrices where the

fronds fell off as the trees advanced in growth. More than two hundred species of ferns have been found in the coal series of Europe. Tree ferns are conspicuous objects in the present flora of Australia; a recent traveler describes one of them which rises to a height of fifty feet, exhibiting a rich crest of fronds, and rivaling even the princely palm-tree in beauty. The ferns are vascular cryptogamous plants, having their organs of fructification on their leaves, as seen on the leaves of the polypody or the brake. Moisture and a warm climate are favorable to their development. The names of the numerous fossil genera of ferns are derived from *pteris*, a fern, to which is prefixed a term indicative of some peculiarity, as *Lonchopteris* (spear leafed fern,) *Sphenopteris* (wedge leafed fern,) Pachypteris (the thick fern,) &c.

Fig. 109.

Sphenopteris.

Fig. 110.

Pecopteris.

Fig. 111.

Pachypteris.

214. Two genera of lofty forest trees appear to have contributed to the formation of coal in large proportion. One of these, the *Lepidodendron*, (*lepidos* a scale; *dendron*, a tree,) was a branching tree forty or fifty feet in length, and four feet in diameter. Narrow sharp

pointed leaves are found attached to it, and scaly cones occur with it which are supposed to be its seeds. This tree is regarded by botanists as intermediate between the Coniferæ and the Lycopodiaceæ or club-mosses.

Fig. 112.

Lepidodendron Sternbergii.

215. The other genus *Sigillaria,* (*Sigillum,* a seal,) was characterized by elegant flutings with rows of symmetrical scars or impressions, disposed with such regularity, as sometimes to have induced the belief that they were engraved stones. These sigilla are scars left in the bark by the separation of the leaves from the stem or the trunk. These trees are usually found in a horizontal position and much flattened by the pressure to which they have been subjected, but are occasionally found erect

Fig. 113.

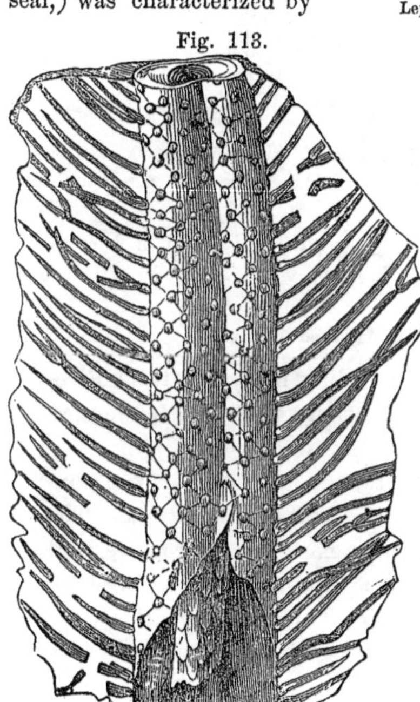

Sigillaria.

with their roots in the soil in the positions in which they grew, and their cylindrical forms are then preserved. The Sigillaria grew sixty feet in length, and five feet in diameter, gradually tapering toward the summit. The bark, which is sometimes an inch thick, is frequently converted into coal, while the interior is sandstone or shale. The fossils called *Stigmariæ* (*Stigma,* a mark,) formerly supposed to constitute a separate genus of plants, are now shown to be the roots of the Sigillaria; a specimen of the Sigillaria has been found in England in an upright position, Fig. 114, with its roots still attached and extending in their natural directions. These stigmariæ are usually found upon the clay beneath the coal beds.

Fig. 114.

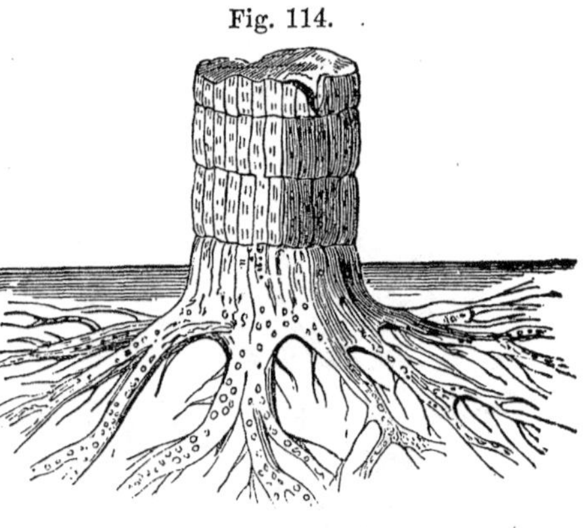

Sigillaria and Stigmaria.

216. The fossil trunks of trees are found in many instances, erect or *inclined* in the coal-bearing strata. The mine of St. Etienne, in France, exhibits numerous vertical stems traversing all its strata appearing like a forest of plants petrified where they grew. In the Craigleith quarry near Edinburgh, a coniferous tree fifty-nine feet long was found lying at an angle of forty degrees, and traversing ten or twelve sandstone strata. This would seem to have

Fig. 115.

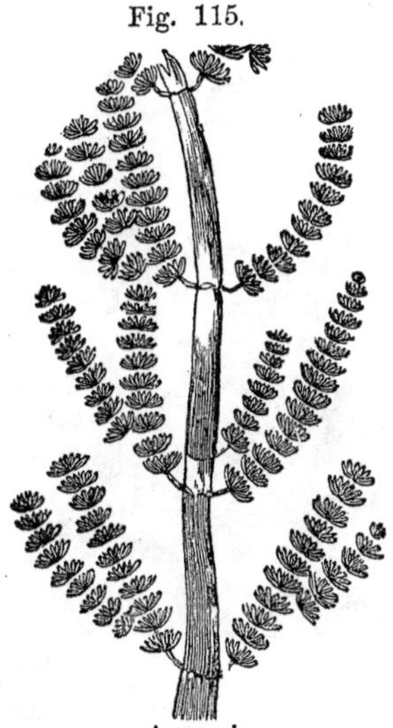

Araucaria.

been due to sudden subsidence or inundation, by which the sand and mud were brought over the trees; or in the cases of inclined stems, the greater weight of the roots may have brought them in that position into the sediment of the water in which they floated, as the snags of the Mississippi are known to work their way many feet into the bed of the river.

217. The origin of the succession of coal-beds with their intervening strata, is not yet satisfactorily ascertained. Some geologists suppose that dense forests and peat-bogs subsiding beneath the water, were covered with the mud and sand, which constitute the shale and sandstone strata over the coal; the land rising again and accumulating vegetable growth, was again submerged—a submergence and elevation occurring for each bed of coal. The number of alternations thus required renders this mode very improbable. Another mode of accounting for the phenomena attributes the coal strata to successive deposits of sand, mud, and vegetables made by rivers in lakes or estuaries, in accordance with the specific gravity of the materials. This mode of action is supposed to be illustrated on a small scale in the deltas of our large rivers, as the Mississippi and Ganges, where immense rafts of vegetable matter are invested by the silt of the rivers, especially during inundations. The perfect state of preservation of many very frail plants of

Fig. 116.

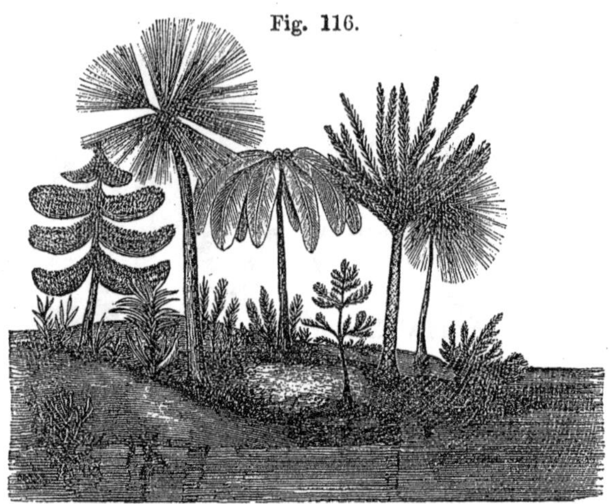

Coal Flora.

the coal strata however, seems hardly consistent with such violent action.

218. The igneous rocks invading the carboniferous strata are trap rock and metaliferous veins. The trap is usually dark colored, presenting itself in dikes, and flattened masses resembling strata, appearing to have been erupted at the bottom of an ocean. The variety of trap called *toadstone*, is abundant in the carboniferous limestone with an aggregate thickness in some places of one hundred and eighty feet. The dislocations of the coal strata

Fig. 117.

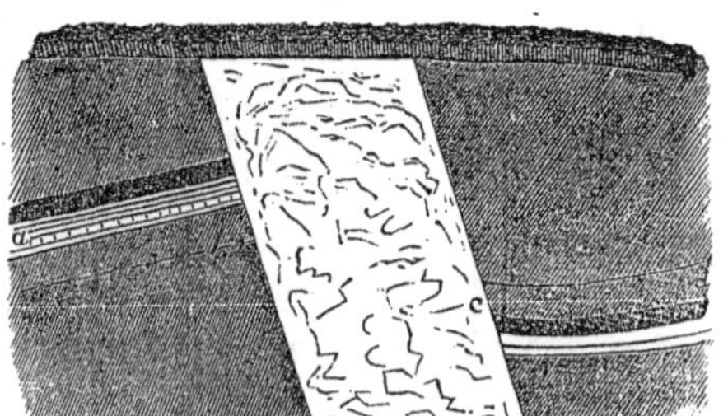

by trap dikes are illustrated by Fig. 117, in which the two parts *a* and *c* of a coal bed are rent asunder by the dike *b*, and the strata on the side *a* cast up two hundred and seventy feet. The texture of the rocks adjacent to the dike has also been changed by its heat. The coal-beds of Belgium have been thrown into a zigzag form over an extensive district, giving rise to an apparent multiplication of the beds; these "troubles," however, sometimes render the coal more accessible.

The carboniferous limestone of this system, is designated *metaliferous:* it is the great repository of lead ore—galena, or sulphuret of lead; iron also abounds in the system.

The rocks of this series often contain *petroleum,* or mineral pitch, which consists of bituminous matter driven out from the coal by subterranean heat; springs or wells of it exist in many localities.

THE PERMIAN SYSTEM.

219. The name *Permian System* has been applied by Mr. Murchison to a fossiliferous series of rocks, which are most perfectly exhibited in Permia extending over a district of 700 miles in length by 400 in breadth, in Russia between the Ural mountains and the River Volga. They occupy an intermediate geological position between the Carboniferous and Triassic systems; and consist of magnesian limestones, marls, red and green sandstones, with beds of gypsum and rock salt. Their fossil contents animal and vegetable resemble those of the carboniferous period. The subdivisions of the system in England are :—

PERMIAN SYSTEM—NINE HUNDRED FEET THICK.

MAGNESIAN LIMESTONE SERIES.	Gray thin-bedded limestone, Red Marl and Gypsum, Magnesian limestone and Magnesian conglomerate.

Lower New Red Sandstone Series.	Marly beds, with thin bands of compact and shelly limestone, Lower new red sandstone.

These rocks are developed in France, and more distinctly in Germany, where they have been thoroughly investigated, and classified. The upper or Magnesian member of the series the Germans designate *Zech-stein*—mine stone—from its containing copper ore; and the lower sandstone, they call *rothe-todte-liegende*—red-dead-lier, because it is of a *red* color, is *dead,* producing no metal, and *underlies* the metaliferous deposit.

220. The lower new red sandstone is closely allied to the upper coal measures, containing some of the same species of extinct vegetables. In mineral composition it differs in being more highly charged with the oxide of iron. The remains of fishes are found in it, and in some of the marly beds in such quantities as to render them bituminous and fetid. The characteristic member of this system is the magnesian limestone, which consists of the carbonate of lime and the carbonate of magnesia in different proportions; the rock is also called *dolomite.* It occurs usually laminated, but presents some remarkable instances of concretionary structure, sometimes in great masses, which resemble piles of cannon balls. The dolomite is comparatively destitute of fossils, while other limestones of the system abound with them. This is the only instance in which an extensive limestone deposit has been so largely charged with magnesia, nor is its origin in this case easily accounted for; some geologists infer from the structure of the rock and other phenomena, that the magnesia, was infused, in the state of liquid or vapor, into the limestone after its deposition.

The prevailing size of the species of fishes found in this formation indicates a diminution from that of earlier periods; but the most remarkable palæontological feature of this epoch is the appearance of reptiles, five species of which have been recognized.

CHAPTER VII.

ROCKS OF THE SECONDARY PERIOD.

221. At the close of the palæozoic epoch, a new era occurs in the history of the globe, marked by violent dislocations of the strata then deposited, depression of extensive portions of the earth's surface beneath the level of the sea, obliteration of races of plants and animals, and the substitution of others in their places. The class of formations which succeeded—the secondary—is well developed in England and on the Continent of Europe in five distinct systems, but is only partially represented in America, Asia, and Australia, as far as is yet ascertained.

THE TRIASSIC SYSTEM.

222. This system is denominated the *Upper New Red Sandstone* in England, but is called the *Trias* on the Continent of Europe, from its appearing in *three* distinct and well marked formations. They have been investigated principally in Germany and France. The following are the subdivisions in Germany, England and France.

TRIAS.

ENGLAND.	GERMANY.	FRANCE.
Saliferous Marls and Sandstones.	Keuper.	Marnes irisees.
Deficient.	Muschelkalk.	Muschelkalk.
Sandstone and quartzose conglomerate.	Bunter Sandstein.	Gres bigarre.

The *Bunter* (variegated) *Sandstein* is a fine-grained sandstone, usually of a red color, but sometimes blue, green, or white, and contains some fossil plants and marine shells.

The *Muschelkalk* (mussel-chalk) is a gray or greenish compact limestone, containing the remains of fishes, radiated animals, and shells in great abundance: some parts of it are highly charged with animal bitumen.

The *Keuper* consists of fine-grained sandstones, and marls of various colors, gray, red, blue, and green, with abundance of rock-salt and gypsum, and few animal remains.

223. The fossil *plants* of this period were *ferns*—in which even the fructification has been preserved—*equiseta*, *coniferæ*, and *cycadeæ*.

Fossil fishes found in the rocks below the trias, are furnished with heterocercal tails; in the triassic period those with homocercal tails were introduced and have continued to predominate to the present day. The *Lily-Encrinite* occurs in the muschelkalk in a beautiful state of preservation. It belongs to the family of Crinoidea, which are found in great numbers in the rocks from the Silurian system upward. These animals had long jointed stems, attached at their bases to the rocks, supporting a cup-shaped cavity, formed by calcareous plates closely fitting each other. This cup contained the viscera of the animal; from its margin rose several arms, subdividing into branches furnished on the inside with numerous *cirri* or feelers. This skeleton was covered by soft parts;

Fig. 118.

Lily Encrinite.

the mouth was situated over the center of the cup. The number of bones belonging to a single individual amounted to one hundred and fifty thousand. These are found sometimes attached, forming a perfect skeleton, but more frequently separate; portions of the stems are called *entrochites, screw-stones, pulley-stones,* and in the north of England, *fairy stones* and *St. Cuthbert's beads.* The stem of the Encrinite is circular, while that of the Pentacrinite is pentagonal. The Cyathocrinite (cup-like Encrinite) was a genus of remarkably light and elegant appearance found principally in the Silurian, Devonian and Carboniferous limestones. The Pentacrinite is exhibited in Figure 106, §209.

Fig. 119.

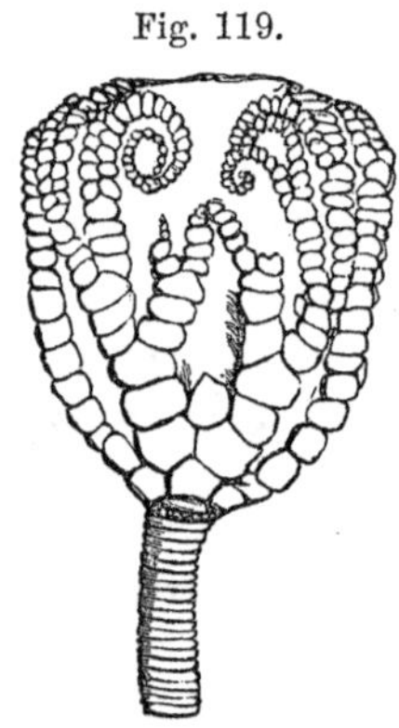

Cyathocrinite.

224. Reptiles are found in such numbers in the secondary rocks, that the period of their deposition has been called the *age of reptiles,* (§152, II.) More than thirty new genera have been added to the class by the study of the fossil forms. Reptiles were formerly described under two orders, distinguished by the presence or absence of external feet. The variety and peculiarities of the fossil species render a new arrangement of the members of the class indispensable. The classification proposed by Professor Owen is admirably adapted to its purpose, including in natural groups all known species, fossil and recent.

Fig. 120.

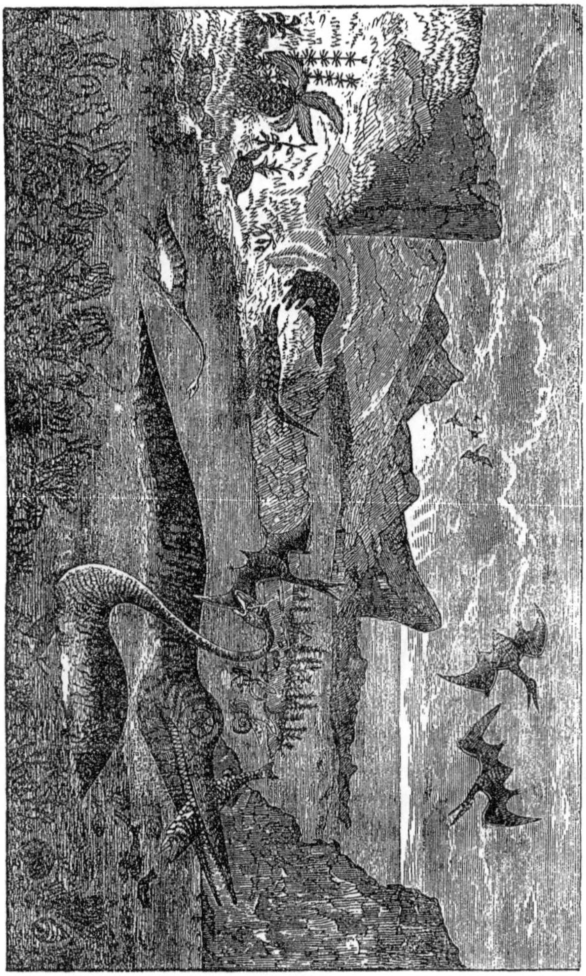

The Age of Reptiles.

	Class III, Reptilia.		*Found fossil in.*
Order	1. Dinosauria	(Land Saurians.)	Oolite, Wealden.
"	2. Enaliosauria	(Marine Saurians.)	Muschelkalk, Lias, Oolite, Wealden, Chalk.
"	3. Crocodilia	(Crocodiles, &c.)	Lias, Oolite, Wealden, Chalk, Tertiary.
"	4. Lacertilia	(Lizards.)	Magnesian Limestone, New Red Sandstone, Chalk.
"	5. Pterosauria	(Flying Saurians.)	Lias, Oolite, Wealden.
"	6. Chelonia	(Tortoises, &c.)	Oolite, New Red Sandstone, Wealden, Tertiary, Chalk.
"	7. Ophidia	(Serpents.)	Tertiary.
"	8. Batrachia	(Frogs, &c.)	New Red Sandstone, Tertiary.

The reptiles of the Triassic period were of the second order, Marine Saurians; of the fourth order, Lizards—the *Rhynchosaurus*, so called from its head resembling the beak of a bird, and the *Dicynodon*, having only two teeth, which were canine; and of the eighth order, Frogs.

225. Peculiar markings, bearing striking resemblance in form and relative positions to tracks made by animals in walking, have long since been observed in the rocks of this period. In 1834, an account was published of some of these observed in the Bunter Sandstein, in Saxony. They resembled the imprint made by a human hand upon any plastic substance, as represented in the adjoining figure. It was proposed to call the animal that made the track, *Cheirotherium* (*Cheir*, the hand; *therion*, beast.) No bones were then identified as belonging to this animal, its existence being inferred from its tracks alone. The dimensions of the tracks are

various, and those made by the hind feet are always much larger than those of the fore feet; in some instances twice as large. The posterior extremities of the animal appear to have been larger and longer than the anterior. Some fragments of skeletons have since been found in the rock with the tracks, which enable Professor Owen to establish a genus of the Batrachian order of reptiles, including several species frequenting the sea shore at the time of deposition of the New Red Sandstone. The jaws were furnished with at least a hundred teeth on each side, diminishing in size from the middle to the extremities; in this respect resembling those of the crocodile. An outline of the animal, as restored by Professor Owen, is presented in Figure 121. As a section of a tooth exhibits, by the

Fig. 121.

Labyrinthodon Pachygnathus.

aid of a microscope, labyrinthine convolutions, the animal is called the *Labyrinthodon.* The tracks of tortoises and crustaceous animals have also been observed in these rocks, and traced in some instances twenty and thirty feet. These fossil foot-prints are called *ichnolites* (*ichne,* track; *lithos,* stone.)

226. Numerous remarkable tridactyle impressions have been discovered in the sandstone of the Connecticut Valley in Massachusetts, which are universally supposed to be

the tracks of biped animals imprinted on the rocks when they were in a soft, forming state. Some specimens exhibit very clearly the character of the foot, its rows of joints, claws and integuments. The best impressions are in fine shale, which is incrusted with micaceous sandstone; the surfaces of the split strata are counterparts of each other; the shales exhibiting the tracks as moulds, and the sandstone as casts in relief. The animals which made these tracks are supposed to have been birds, though none of their bones have yet been found in these strata. President Hitchcock distinguishes more than thirty species by differences in the tracks. They are of various sizes; some as small as our sparrows, while some made a track fifteen inches long, exceeding by five inches the track of the African ostrich. The length of the stride of these largest species was from four to six feet. The doubts formerly entertained respecting these *Ornithicnites* have been dissipated by the discovery, in the alluvial deposits of New Zealand, of the skeletons of wingless birds as large as those to which the tracks in the New Red Sandstone are assigned, (§301.) These bird tracks have been presented in Figure 69, § 126.

227. The rocks of this period seem to have been especially fitted to perpetuate impressions: the ripple marks on them are very distinct; but, what is more remarkable, their surfaces often present a pitted appearance, dotted with little hemispherical eminences or depressions which are attributed to rain-drops. These are sometimes elongated, as if the drops were driven by the wind in an oblique direction. President Hitchcock remarks, "It is a most interesting thought, that while millions of men, who have striven hard to transmit some trace of their existence to

future generations, have sunk into utter oblivion, the simple footsteps of animals that existed thousands, nay, tens of thousands of years ago, should remain as fresh and distinct as if yesterday impressed: even though nearly every other vestige of their existence has vanished: nay, still more strange is it that even the pattering of a shower at that distant period, should have left marks equally distinct, and registered with infallible certainty, the direction of the wind!"

Fig. 122.

A Footprint and impressions of Rain-drops.

228. The origin of *rock salt* in masses, is usually ascribed to the evaporation of the water in brine springs, or portions of the sea isolated and exposed to a high temperature. The salt is in such circumstances precipitated, and accumulates, as is seen in the Great Salt Lake, the Dead and Caspian seas, and the lakes of northern Africa and Arabia. Some phenomena of salt beds, however, indicate a connection with volcanic agency. Salt is not confined to the saliferous deposits of the New Red Sandstone, but is found in the carboniferous, oolitic, and tertiary rocks.

THE LIASSIC SYSTEM.

229. The *Lias* consists of strata mostly argillaceous; but having a large quantity of calcareous matter also, it embráces many bands of argillaceous limestones. Some beds of marly sandstones alternate with the clays and limestones. The term lias is probably a corruption of *layers*, alluding to the riband-like appearance which a section of the rocks of the system presents. This system of rocks is developed in England, and on the continent of Europe, in France, Switzerland, and Germany; it is supposed to exist also in Asia and South America, but it is doubtful whether there are any of its deposits in North America. Its subdivisions are:

LIAS, 600 feet thick.	**Upper Lias** or **Alum Shale**—soft shales extremely fossiliferous, jet and bituminized wood, hard shales.
	Marlstone—colored calcareous clays and sands, impure limestone.
	Lower Lias Shale—calcareous flagstones, laminated dark-colored shales.
	Lower Lias—a thin bed, of argillaceous limestone, containing numerous fossil fishes; sometimes passing into sandstone.

230. The Lias formation is surpassingly rich in fossils, embracing representatives of all the great natural families. The Corals are few and small, but the Crinoidea were exceedingly abundant; whole beds are made up of *Pentacrinites*, so perfectly preserved that their minutest anatomical structure is discernible. These animals are supposed to have had the power of attaching and detaching themselves at will; large masses of them are found attached to the fossil wood of this period. Bivalve shell-fish

Fig. 123.

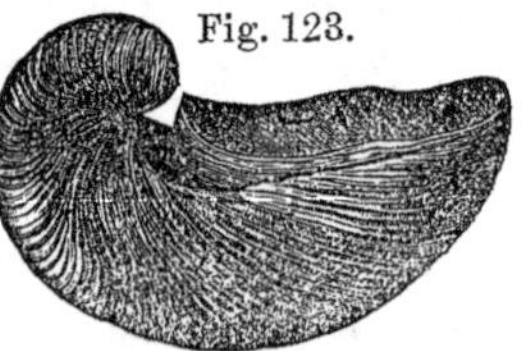

Gryphæa Incurva.

are very numerous, some of which are characteristic of the formation, and others of particular beds; one of the marly limestone beds is called the *Gryphite Limestone,* from the great abundance of the peculiar fossil *Gryphæa incurva,* Fig. 123. The *Terebratulæ* were still continued, and the genus *Spirifer,* which however terminates its existence in this formation, being found in no newer rock.

231. The *Ammonite,* though early introduced in the strata, appears at this period to have been greatly expanded; more than seventy species of the genus have been found in the British Lias. It was a univalve, many chambered shell, whose inhabitant belonged to the order Cephalopoda. It was allied to the Nautilus, which resides in the outer apartment of its shell, communicating with the interior chambers by means of a tube—siphuncle—and is supposed to rise and fall through great depths of water, diminishing or increasing its specific gravity by withdrawing or injecting water into the internal cavity. As the shells of the Ammonite were thin, they were often strengthened by ribs and tubercles. They vary in size from half an inch to four feet in diameter.

Fig. 124.

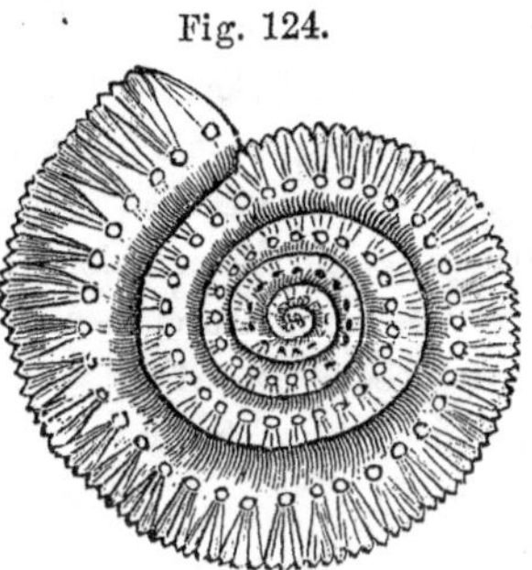

Ammonite.

232. The *Belemnite,* (so called from its resemblance to a dart,) occurs very perfectly preserved in the Lias, Oolitic, and Chalk rocks. It consists of a calcareous, cylindrical or conical shell, pointed at one extremity, and terminating at the other in thin conical chambers. Some specimens are small, amber-colored, and transparent; while

others are opake, ten or twelve inches long, and four inches in circumference. The animal belonged to the Cephalopoda, was allied to the Cuttlefish, and was furnished with an ink-bag. The shell, (Fig. 125,) was, like the *pen* of the Cuttlefish, internal. The ink-bag and its contents have been found fossil, and a pigment, *india ink*, has been prepared from them possessing all the properties of the ink prepared from the recent Sepia. The soft parts of the animal are in many instances so well preserved, that its form and structure are discernible. Its eyes were large and prominent, and its horny beak powerful. Its habits were probably, like those of the family of the present day, extremely voracious.

Fig. 125.

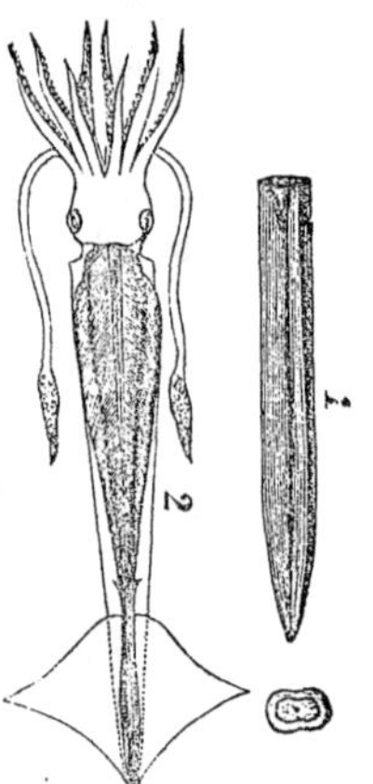

Belemnite.

233. Fishes and Reptiles were the representatives of the Vertebrata at this period; of the former more than sixty species have been discovered in the Lias of England, belonging almost exclusively to the Ganoid and Placoid orders, and all of them extinct. Numerous fossil spines, or bony rays of the dorsal fins of extinct cartilaginous fishes are found in the Lias rocks, detached and isolated; they are called *Ichthyodorulites*. They were not articulated to the vertebræ, but deeply implanted in the flesh. Some of the existing species of shark are provided with the same bony ray, giving to the dorsal fin greater force and precision in its motions.

234. The Reptiles of the Liassic period belong to the

marine order—Enaliosauria—and are principally confined to two genera, *Ichthyosaurus* and *Plesiosaurus*.

The Ichthyosaurus, or fish-lizard—so called because it combines in its structure the features of the fish with those of the reptile—resembled the Grampus or Porpoise in figure, with a much larger head and a long tail expanded

Fig. 126.

Ichthyosaurus.

vertically, as are those of true fishes. It was covered with a tough skin, like that of the whale, destitute of scales. The orbit of the eye was unusually large—in some specimens eighteen inches in diameter—and enclosed a series of bony plates, enabling the eye to adjust itself to different distances, and varying intensities of light; an analogous

Fig. 127.

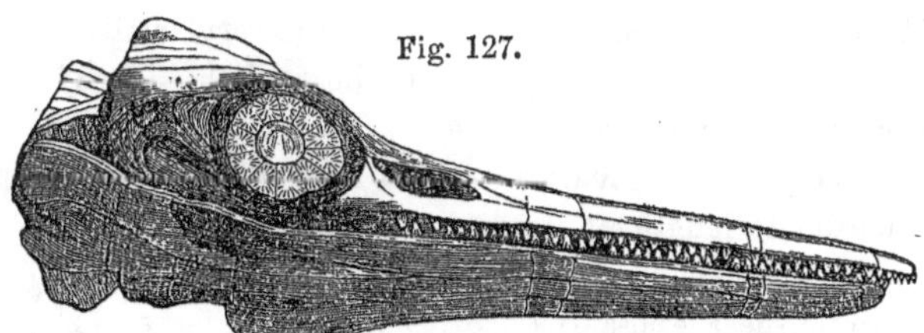

Head of Ichthyosaurus.

structure is exhibited by the eyes of turtles, lizards, and especially of the owl. The teeth were conical, like those of the crocodile, and very numerous, amounting in some species to two hundred; in some large individuals the opening of the jaws exceeded seven feet. Its organs of locomotion were four paddles composed of numerous bones, surrounded by rays of true fins; the anterior pair were based upon a firm breast bone, which enabled the animal to give rapidity and precision to its motions. The form and arrangement of its vertebræ, which were in some species one hundred and forty in number, were like those of fishes. Some individuals attained a length of forty feet. They were carnivorous, as is shown by their teeth, as well as by the contents of their stomachs which have been found in a few instances preserved.

235. The Plesiosaurus (*plesion*, near; *saurus*, reptile,) approached more nearly the true reptiles in structure. Its

Fig. 128.

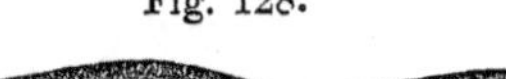

Plesiosaurus.

head, which resembled that of a lizard, was very small; its paddles were relatively longer than those of the Ichthyosaurus, and its tail was like that of the crocodile. The most remarkable peculiarity exhibited by this animal was the inordinate length of its neck. Animals have usually not more than five cervical vertebræ; the swan has twenty-four, while the Plesiosaurus had from thirty to forty. The animal seems to have been fitted for rapid motion. It attained a large size; a perfect specimen has been obtained seventeen feet long, and fragments of larger skeletons have been found. Numerous species of the Ichthyosaurus and Plesiosaurus have been found in the rocks from the muschelkalk to the chalk.

236. Beds of *jet*, resembling cannel-coal but harder and more lustrous, are found in the Upper Lias, together with large fragments of bituminized wood of coniferous trees and Cycadeæ.

THE OOLITIC SYSTEM.

237. The Oolite is a marine formation of great extent and thickness, consisting principally of alternations of limestones and clays, embracing vegetable remains, and very numerous corals, shells, fishes and reptiles; insects and mammalia also are found in it. The Oolite derives its name (*Oon*, an egg; *lithos*, a stone,) from the fact that some of its beds consist of calcareous globules which resemble the roe of a fish; but this structure is not strictly characteristic of it. This system is developed in Great Britain and on the Continent of Europe, in France and Germany, (called the Jurassic series,) in Russia, India and America. Its subdivisions in England are:

UPPER OOLITE.	**Portland Beds**—limestone of an oolitic structure, abounding with marine shells, ammonites, &c. **Kimmeridge Clay**—blue and yellow clay with septaria, gypsum, marine shells and beds of lignite or Kimmeridge coal.
MIDDLE OOLITE.	**Coral Oolite** or **Coral Rag**—coarse limestone full of corals and shells—beds of yellow sands and calcareous grits. **Oxford Clay**—dark blue clay with septaria and numerous fossils—beds of calcareous sandstone.

Lower Oolite	**Cornbrash**—coarse gray limestone separated by beds of clay. **Forest Marble**—sand, with thin fissile limestone and coarse shelly oolite. **Great Oolite**—oolitic limestone and freestone, with beds of clay, and remains of reptiles, corals, &c. **Stonesfield Slate**—oolitic limestone, with remains of land animals, plants, insects, and mammalia. **Fuller's Earth Beds**—marls and clays containing fuller's earth. **Inferior Oolite**—coarse limestone, ferruginous sand with shells.

238. The alternations of limestones with clays and shales, displayed by the Oolitic series, give rise to peculiar features of outline in parts of England and France. Three successive ridges are the hard calcareous beds of the Lower, Middle and Upper Oolites, while the intervening plains and vallies consist of clays and shales.

239. The organic remains of the Oolite are numerous and varied, as might be expected from the variety of mineral constitution of the strata.

The corals occur chiefly in the Middle Oolite; the coral rag consists of an extensive coral reef thirty or forty feet thick, composed mostly of a few genera, some of which, as the Caryophyllia, are represented by species in the present seas.

Among the Crinoidea, the Pear Encrinite—Apiocrinites—similar in structure to the Lily Encrinite, grew abundantly, attached to rocks.

The Sea Urchins—*Echini*—were also numerous.

Of the mollusca, the bivalve acephalous were very abundant; some beds are almost entirely made up of their remains. The Trigonia Costata, Fig. 129, abounded in the Oxford Clay. Ammonites and Belemnites were as common as in the Liassic period.

Fig. 129.

Trigonia Costata.

Crustacea appear in the Oolite with forms closely allied to those of the present day. Figure 130 presents an animal resembling the Lobster, found in the Oolitic clay.

Fig. 130.

Fossil Crustacean.

Insects also are found, some of which very closely resembling the common Dragon-fly, are so perfectly preserved that the most delicate markings, and the nerves of the wings are distinctly discernible.

240. Of the vertebrate animals, fishes are found in abundance throughout the series, belonging mostly to the ganoid order. The Ichthyosaurus and Plesiosaurus, so abundant in the Lias, extend through the Oolite; and other saurians of gigantic size are found in these strata—*Megalosaurus, Teleosaurus, Steneosaurus* and *Cetiosaurus.* But

the most anomalous saurian of this period, was the Pterodactyle, (*Pteron*, a wing; *dactulos* a finger,) an animal, of which Cuvier remarks, "It is undoubtedly the most extraordinary of all the beings of whose former existence a knowledge is granted to us, and that which, if seen alive, would appear most unlike anything that exists in the present world."

Fig. 131.

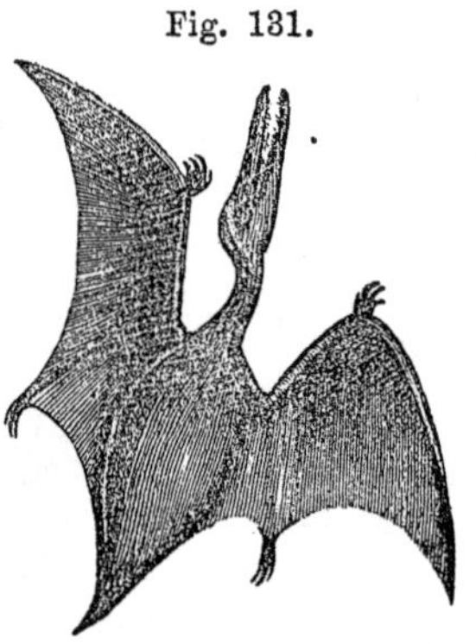

Pterodactyle.

The Pterodactyle bore some resemblance to the bat, but its mouth was very unlike that of the bat, having the jaws and teeth of a reptile. It was furnished with powerful membranous wings, enabling it to fly with ease and rapidity, and its feet were not so involved in the membranous expanse of the wings as to prevent walking. Its eyes were very large.

241. The Stonesfield slate furnishes the only specimens of fossil mammalia yet discovered in the secondary rocks. Parts of the skeletons and the teeth of two genera of the marsupial order of mammiferous quadrupeds have been found in this rock.

Fig. 132.

Jaw and Teeth of the Phascolotherium.

The Phascolotherium was allied to the Opossum and Kangaroo; and the organic remains of the Oolite generally resemble the present fauna and flora of Australia.

242. The *plants* of the Oolitic period consist of coniferæ, liliaceæ, palms and ferns; but the Cycadeæ, which were introduced in small numbers in the Carboniferous period, are here found abundant. They are intermediate in character, between the Coniferæ, palms and ferns. Some of them are short, stout, cylindrical, scaly stems, surmounted by a tuft of elegant leaves resembling a pine apple, while

Fig. 133.

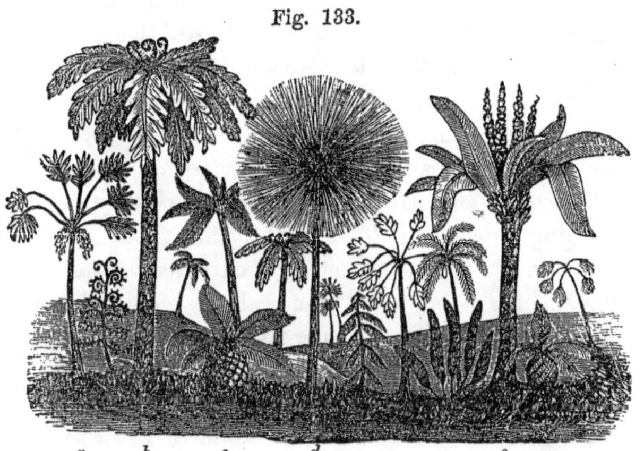

a Palm. *b* Arborescent fern. *c* Cycas. *d* Pandanus. *e* Zamia.

others have trunks thirty feet in length. The Cycadeæ still grow in tropical regions. Valuable beds of lignite and bituminous coal are found in the Oolite, as the Kimmeridge coal, Brora coal and the Yorkshire Oolitic coal, in England; but the most important coal beds of the Secondary Period, hitherto discovered, are those of Richmond, Virginia, which cover one hundred and eighty-five square miles, varying in thickness from eleven to forty feet. There are three beds exhibiting the same phenomena that are observed in the older coal beds. These beds are extensively wrought, the coal yielding abundant pure gas. The leaves of cycadeous plants are abundant in these beds.

THE WEALDEN SYSTEM.

243. Hitherto, with the exception of a very few thin strata, the whole series of deposited rocks have been of marine origin, as is shown by their mineral constitution and organic contents. The Wealden System exhibits clear indications of being fresh-water deposits, containing numerous remains of land plants and animals. This formation consists of a thick series of sandy beds resting on an imperfect limestone, and covered by a bed of clay; the whole containing fresh-water shells, land plants, fishes and reptiles, which indicate the action of river currents, but not the attrition of the ocean upon them. It resembles the deltas forming at the mouths of large rivers at the present day. The Wealden appears in England, Scotland, France and Germany. The term Wealden is derived from *wealds* or *walds*, the ancient word for woods; because these rocks form a tract of land occupied by the forests of Kent and Sussex, England. The order of the beds as observed in England, is

WEALDEN SERIES, 1000 feet thick.	**Weald Clay,** with subordinate limestone and sand—fossil fresh-water shells, almost entirely composing some beds.
	Hastings Sand, including the beds of Tilgate Forest—bluish gray and ferruginous sandstones, conglomerate, lignite, fresh-water mollusca.
	Purbeck Strata—compact limestone alternating with clay, resting upon the Portland beds of the Oolite.

Fig. 134.

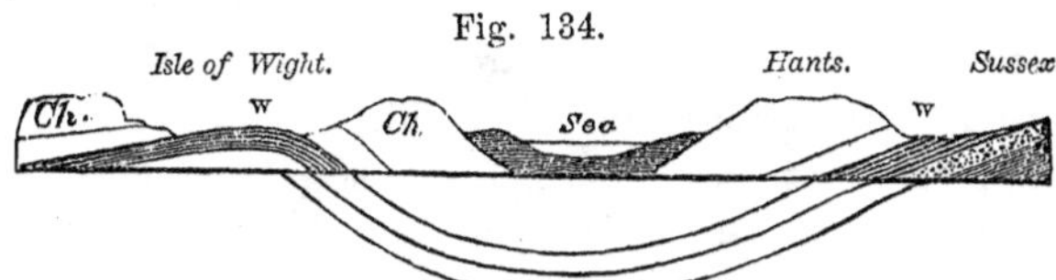

Section of the Chalk and Wealden. W—Wealden.

The relative positions of the Chalk and Wealden are presented in Fig. 134, in which the Wealden invariably underlies the Chalk. The Oolitic Series is made, by some writers, to embrace these rocks, and by others they are assigned to the Chalk; but their organic characters especially entitle them to rank as a distinct system.

244. While some of the organic remains that occurred in the Oolite are found in this formation, most of the Wealden fossils are peculiar, being almost exclusively fresh-water, or terrestrial in their origin. The number of species is not large, but the abundance of remains of a species surpasses what has hitherto been witnessed. The remains of *shellfish* and *crustaceans* are in some of the beds exceedingly abundant, and indicate an approximation, in form, to the animals of those classes now existing. Beetles and other *insects* are often found in these strata.

245. Among vertebrate animals there are the remains of several species of *fish* some of which are allied to the *lepidotus*, bony pike of the North American lakes; but the *reptiles* furnish the most remarkable fossils in the Wealden.

To the Plesiosaurus, Megalosaurus and Cetiosaurus found in the Oolite, are added five genera of Saurians peculiar to the Wealden; among which are the Hylæosaurus, and Iguanodon belonging to the order *Dinosauria*.

The *Hylæosaurus*, or *Weald Lizard*, was a land reptile, twenty feet in length, covered with elliptical scales.

The *Iguanodon*, so called from the resemblance of its teeth to those of the Iguana, a lizard found in the West Indies, was an herbivorous reptile, about thirty feet long and of colossal proportions, having a thigh-bone larger than that of the Elephant. It was furnished with a short horn as are some species of Iguana.

The remains of a large fresh-water turtle and other Chelonian animals, and fragments of bones belonging to an extinct species of wading bird are found in the Wealden, but no traces of mammalia have been detected.

246. The general character of the Wealden vegetation is the same as that of the Oolite, with the addition of a few peculiar species.

Fig. 135.

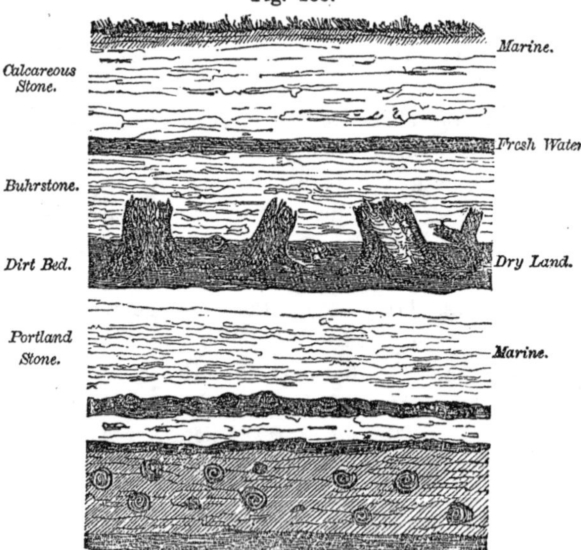

Subterranean Forest—Portland Dirt-bed.

Between the marine beds of the Oolite and the fresh-water deposits of the Wealden intervenes a bed eighteen inches thick, of black vegetable soil, called the "Dirt-bed." In it are found numerous trunks and stumps of coniferous trees, fossilized where they grew, with their roots still immersed in the ancient soil. Here are distinct intimations of elevation of the Portland stone to the surface of the sea in which it was deposited, so as to constitute the soil for the growth of trees; the growth of vegetation so as to produce eighteen inches of earthy lignite; the subsequent depression so gentle as not to remove the trees from their position; and finally the deposition of fresh-water and marine strata over the whole.

Coal beds from one to three feet thick are found in the Wealden of Germany.

THE CHALK OR CRETACEOUS SYSTEM.

247. The Cretaceous group is a marine formation comprising limestones, marls, clays and sands, abounding with remains of marine origin. The name of the system is derived from its leading member, Chalk, (*creta.*) This formation is extensively developed in England, Ireland, France, Germany, Russia, Spain, Italy, Northeastern Africa, India, South America and the United States. Its subdivisions in England are

CRETACEOUS SYSTEM—ONE THOUSAND FEET THICK.

CHALK FORMATION.	**Upper Chalk**, soft, white—with flints.
	Lower Chalk, hard white—few or no flints
	Chalk Marl—marl with silicate of iron.
GREENSAND FORMATION.	**Upper Greensand**—sand, chert, ocnre.
	Gault—blue marl with ferruginous nodules.
	Lower Greensand with occasional limestone.

The *Greensands* consist of silicious sand mixed with the silicate of iron—chlorite—which gives the green color; the color is however often yellowish, or gray, from admixture with the oxide of iron. These sands alternate with beds of clay and limestone, and contain frequently fuller's earth and chert.

Gault is a provincial term used in England to designate the brick-clay found in this system. It is a stiff blue clay containing iron-pyrites, and in some portions, the green silicate of iron. It does not occur in the cretaceous system of the United States.

The *Chalk* is a familiar, earthy form of the carbonate of lime, sometimes soft, but frequently sufficiently solid for building purposes. Its stratification is often indistinct, but recognizable by the alternating layers of flint in it.

Fig. 136.

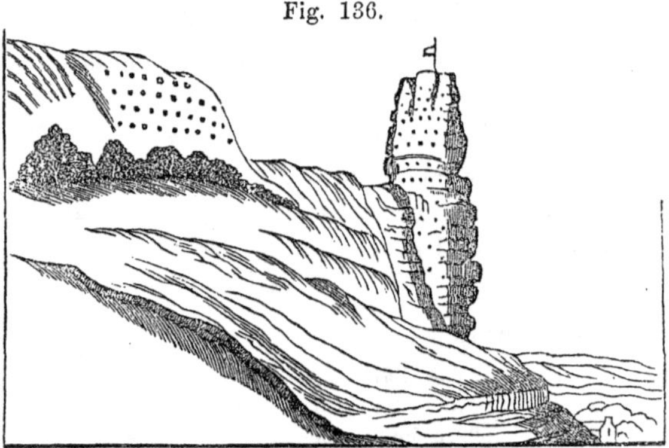

Chalk with Flints—An Outlier.

The Upper Chalk is remarkable for its numerous bands of dark colored flints; bands from three to six inches in thickness, and from two to four feet apart. These flints have very often an organic body as a nucleus about which they have accumulated, and whose form they have preserved. In the Lower Chalk the flints are rarely found, but the silica is diffused through the mass. The lowest portion is largely mixed with clay constituting the chalk marl, which sometimes is green from the presence of the silicate of iron, or red from the oxide of iron.

248. The Chalk fossils are numerous, almost exclusively marine, and appear to have been deposited in deep water.

Among the radiated animals, the Encrinites had dwindled to a few species of small size and diminished beauty; one form of them in this period was the *Marsupite,* which resembled the lily encrinite deprived of its stem. Sea Urchins—echini—were very abundant, accompanied by crabs and lobsters whose organization closely resembled that of the members of those tribes of the present day. Bivalve and univalve shells are found abundantly. The cephalopodous animals to which the ammonite belongs, exhibited a great diversity of forms; curved in successive whorls like the ammonite, but with the whorls slightly remote from each other, as the *Crioceratite;* hook-shaped, as the *Hamite;* or straight like the Orthoceratite, as the *Baculite* and *Turrilite.*

249. Vast numbers of small animals existed at this period living in shells formed of numerous chambers or compartments, belonging to a group which is still common in the ocean—the *Foraminifera.* These animals occupy all the chambers of their habitation, and not the outer one only as does the nautilus. Their shells are sometimes flat and disc-like, resembling a piece of money: hence we have the *Nummulite,* from *nummus,* a piece of money, Fig. 137. Though of small size they often constitute almost the entire mass

Fig. 137.

Nummulite.

of mountains. Nummulitic limestones were used in the construction of the Sphinx and pyramids of Egypt.

Infusoria also abounded during the deposition of the chalk; but their beauty and perfect state of preservation in the flints, can be discerned only by the aid of powerful microscopes; a large proportion of the chalk itself appears to be made up of the fragmentary skeletons of these minute animals. The shapes of the skeletons of these animalcules are various; some consist of parallel tubes arranged transversely upon long fragile ribands;

Fig. 138.

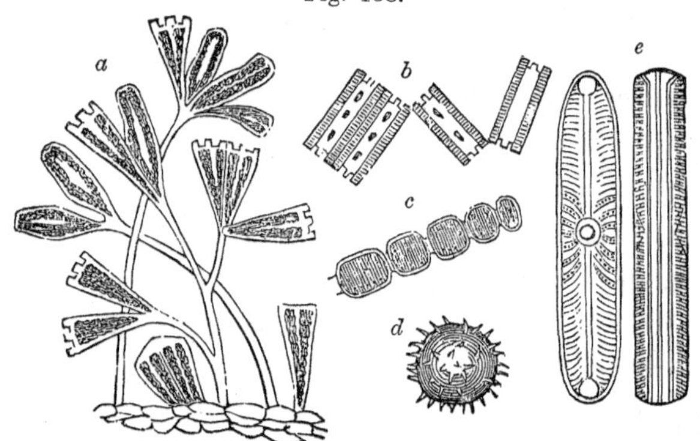

Fossil Infusoria.

a Gomphonema. *b* Bacillaria. *c* Gaillonella. *d* Xanthidium. *e* Navicula.

others are oblong, or globular, with their surfaces studded with hair-like appendages; while others still are in ramose groups like polyps. Many flint nodules are almost entirely composed of the silicious skeletons of these infinitesimal forms of animal life. Fig. 139 presents one of the fossil infusoria, closely resembling the living Xanthidia. It is here magnified 500 times linear. Fig. 140 represents the cell of a zoophyte in flint, with the polyp protruded, having been entombed in the mineral matter while alive and active. This also is magnified 500 times linear. The Gaillonellæ are another family of infusoria found in the flint, so minute that forty-one thousand millions of their skeletons occupy but a cubic inch. Their cylindrical shields are arranged in a long series like a tubular chain. They multiply by self division with such rapidity that a single individual may produce one hundred and forty millions of millions in twenty-four hours.*

Fig. 139.

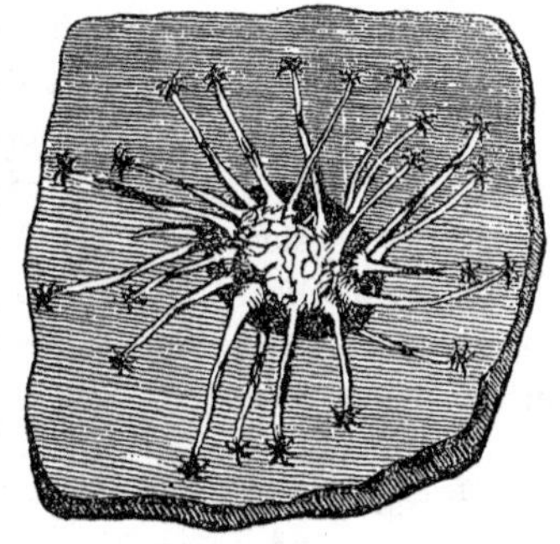

Xanthidium Tubiforme.

Fig. 140.

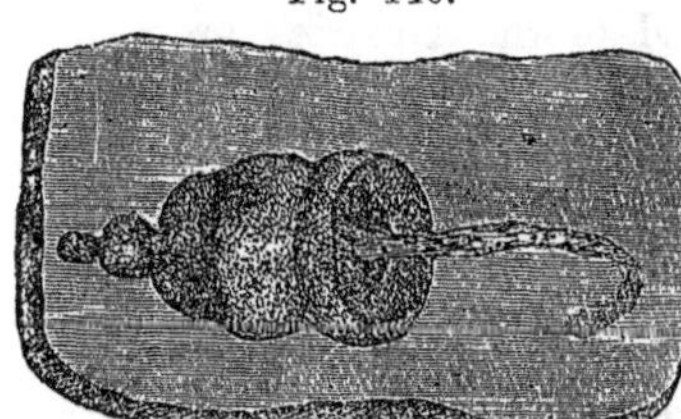

Zoophyte in Flint.

250. Of the vertebrate animals, the chalk deposits include some sauroid *fishes*, among which the *Macropoma*, a voracious fish about two feet long, has been found so well preserved that its gills and entire stomach with its contents are presented to inspection. The teeth and bony spines of sharks are abundant, and the other two orders of fishes—the ctenoids and cycloids—which so greatly pre-

* Mantell's Wonders of Geology, I. p. 565.

dominate in the present seas, are also introduced in the cretaceous rocks.

The *reptilian* remains of the Cretaceous series compared with the three preceding systems are few, embracing, so far as has been discovered, but one or two forms that were not known before. The Ichthyosaurus and Pterodactyle of the previous system were continued in this, and a new gigantic marine lizard, nearly allied to the Monitor of the present period, was introduced, called the Mosasaurus—the saurian of the Meuse, on whose banks its fossil remains have been found. It attained the length of twenty-five feet, and appears to have been the last gigantic form of marine reptiles introduced, before their obliteration at the close of the secondary period. Its teeth are found in the cretaceous deposits of the United States. Several species of marine turtles occur in the cretaceous rocks.

The bones of *birds* allied to the Albatross, are recognized among the fossils of the chalk, but no trace of any land quadruped, or any member of the whale tribe.

251. The *sponge,* a peculiar kind of organism consisting of horny fibres, often enclosing spiculæ of flint or carbonate of lime, has very frequently served as a nucleus about which the silicious or flinty matter, diffused in the sea at the period of the chalk deposits, accumulated, and the forms of the sponge have thus been perfectly preserved in flint, in chalk and sandstones. Carbonized sea-weeds, drift-wood, and beds of lignite occur in the chalk group of rocks.

252. The *origin* of chalk, so different in its texture and appearance from the other limestones, with its included flints, has been the subject of much discussion. It has been supposed by some to be a chemical precipitate: "but there appears no evidence," says Mr. Brande, "of its having

been deposited from chemical solution; on the other hand, it bears marks of a mechanical deposit, as if from water loaded with it in fine division. And upon this principle some gleam of light may perhaps be thrown upon the enigmatical appearance of the flints; for it is found that if finely divided silica be mixed with other earthy bodies, and the whole diffused through water, the grains of silica have, under certain circumstances, a tendency to aggregate into small nodules; and in chalk some grains of silica are discoverable." The flints occur in nodular masses of irregular forms, varying from less than an inch to more than a yard in circumference; they are quite insulated from each other, each one being entirely enveloped by the chalk. As they almost invariably include some organic body, they appear to be silicious aggregations about nuclei, as septaria are aggregations of calcareous and argillaceous matter. It has before been shown (§ 66,) that silica is soluble in water, as it occurs in the Geysers of Iceland. The chalk resembles the calcareous mud that accumulates in the lagoons of coral islets, produced by the ocean wearing corals and shells to a fine powder. (§ 79.)

253. The *igneous* rocks associated with the chalk and other members of the Secondary series, are basalt and other forms of trap. The basalt of the Giant's Causeway breaks through and overlies the chalk in the north of Ireland, the heat of the igneous rock having rendered the chalk hard and crystalline like primary marble.

CHAPTER VIII.

ROCKS OF THE TERTIARY PERIOD.

254. THE Tertiary embraces the strata of sand, clay, and limestone, which lie between the chalk and the superficial deposits designated Drift and Alluvium, with which they were confounded previous to the labors of Cuvier and Brogniart in A. D. 1810. They differ from the Secondary rocks in mineral constitution and especially in their organic contents; but are not very distinctly bounded at their upper limit, frequently merging in the recent deposits. In general they are not so firmly aggregated nor so thick as the Secondary rocks, and are remarkable for exhibiting frequent alternations of marine and fresh-water formations in the same localities. These beds very frequently occur in limited and detached basins, as if deposited in shallow lakes or estuaries; they are however often continuous over very extensive tracts of country, as from Martha's Vineyard southward along the whole extent of the Atlantic coast of the United States. The order of succession of strata is not so uniform as in the older rocks; and hence arises the difficulty of framing a description that will apply to them in all localities. Their alternations have, however, been accurately ascertained, especially in the Paris and London basins.

255 The *sands* and *clays* predominate in the Tertiary series; their colors are various, depending principally upon

the colored compounds of iron, and they are not sufficiently indurated to form firm sandstones and shales. The calcareous strata are more varied in structure and appearance, consisting of soft marls filled with shells—of rough masses of coral—of fresh-water beds of hard limestone—and of marine limestone of coarse sandy texture. They also contain beds of silica, buhrstone, gypsum, and rock salt.

256. The Tertiary strata occupy a large portion of the surface of Europe, conforming in a remarkable degree to the present outline of the sea; so that if the continent were depressed a few hundred feet, the sea would cover them, indicating that no essential change has occurred in the figure of the land since that period. The Tertiary series has been recognized also, in Northern Africa, in Asia, in North America, and in South America on both sides of the Andes.

257. These rocks were, at first, *classified* in accordance with their alternations as marine and fresh-water deposits; but these alternations are found not to be uniform. The classification generally adopted is that of MM. Lyell and Deshayes, based upon the relative proportion of shells specifically identical with those occurring at the present time. Shells are selected as the test because they are more generally and uniformly diffused in strata than the remains of any other class of animals.

Eocene, (*Eos*, the dawn; and *kainos*, recent,) the dawn of the present period with its races of plants and animals; contains less than five per cent. of living species, in its fossil contents.

Meiocene, (*Meion*, less; *kainos*, recent,) though containing more living species than the Eocene, they are still less than half; its per centage of recent species is eighteen.

PLEIOCENE, (*Pleion*, more; *kainos*, recent,) a majority of whose fossils belong to existing species.

PLEISTOCENE, (*Pleistos*, most; *kainos*, recent,) a still nearer approximation to the present period, having ninety-five per cent. of recent shells.

THE EOCENE OR OLDER TERTIARY.

258. The Eocene deposits are the most distinct of the series, and occur well developed in France and England. They, however, vary much in local character. The beds of the Paris and London basins are of the same age, but differ greatly in mineral constitution; in the London group those of a mechanical origin—sands and clays almost exclusively prevail; while beds of limestone, silica, and gypsum abound in the Paris basin.

The following tabular arrangement exhibits the order of the beds in the two basins:

PARIS BASIN.	LONDON BASIN.
Upper Fresh-water group—marls, sands, shelly limestone, buhr limestone.	
Upper Marine—sands and marls with thin layers of limestone.	
Lower Fresh-water—marls, silicious limestones, gypsum with numerous bones of animals.	*Bagshot Sand*—sand containing few fossils.
Lower Marine—coarse sandy limestone (*calcaire grossier*) calcareous marls and green sand.	*London Clay*—clay, blackish or gray, containing septaria and ferruginous nodules, with numerous organic remains.
Plastic Clay—blue plastic clay, sand and pebbles, beds of lignite.	*Plastic Clay and Sands*—sands and sandy clay, with shells; beds of lignite, flint pebbles.

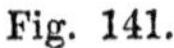

Fig. 141.

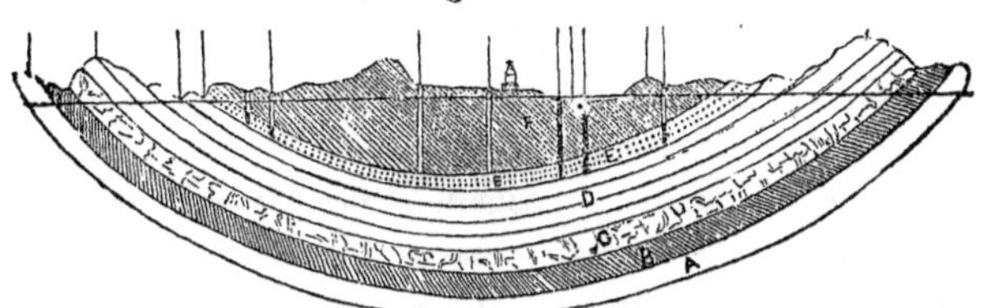

Section of the London Basin.

F represents the London Clay; E the plastic clay; D the Chalk; C the Wealden; B the Oolite; the Bagshot sand caps and overlies the London Clay. The position of the City of London is indicated by the building.

259. There are numerous other localities of the Eocene Tertiary, as in Southern France, Spain, Belgium, Italy, Greece, Asia Minor, Egypt, India and North and South America.

260. The fossils of this group are very numerous and important, indicating the condition of the earth at the time of their deposition. The general character of the flora and fauna was like that within the tropics of the present day. The species are with very few exceptions extinct, and seem to have been almost exclusively confined to this formation. More than a hundred species of *plants* have been distinguished, many of which are dicotyledonous, and resemble intertropical plants of the present day; a great number of their seeds and fruits have been found in the London clay beds of the Isle of Sheppey. Various fossil resins have been discovered in these beds, the most important of which is *amber*, occurring in nodules from the size of a nut to that of a man's head; one specimen weighs 18 lbs. Specimens frequently contain insects whose positions and detached legs and wings indicate that they struggled after

they were involved in the resin which was at that time liquid.

261. A large number of shells, both univalve and bivalve, occur in the Eocene: the nautilus was retained, but the ammonites and belemnites had become extinct. More than two hundred species of *Cerithium* (an elongated shell resembling Fig. 142) living in brackish water, are found in the European Eocene; and the *nummulites*, Fig. 137, are so abundant in some localities as to make up almost the entire mass of rock. The remains of *crabs* are abundant in these deposits. About one hundred species of *fish* have been recognized. Their remains at Monte Bolca in Northern Italy, and at Mount Lebanon are very numerous; the immense quantities which have been found in a beautiful state of preservation at the former locality, have been thought to indicate that the limestone in which they are imbedded was erupted into the ocean in a fluid state, by volcanic action, suddenly suffocating and enveloping them in the calcareous mass. The *species* of fish found in the Eocene are all extinct, and one third of them belong to extinct *genera*; but many of them are closely allied to existing species. The ganoid and placoid orders, characterized by their covering of pointed bony plates and enamelled scales, which abounded in the older rocks, had become relatively rare; the placoids, however, had representatives in the sharks and rays, whose teeth are found in abundance. The ctenoids and cycloids, which had been introduced just before the close of the

Fig. 142.

Fig. 143.

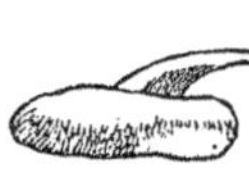

Placoid Scales.

secondary period, now predominated. The ctenoid order is characterized by scales having the posterior margins finely pectinated —divided into little teeth like a comb; these scales are imbricated, each lamina being smaller than the one beneath it, and pectinated as represented in Fig. 144. This order is represented by the perch family. The cycloid order is characterized by circular imbricate scales with smooth margins, of which the mackerel furnishes an example. More than three-fourths of the species of fishes now living belong to the ctenoid and cycloid orders.

Fig. 144.

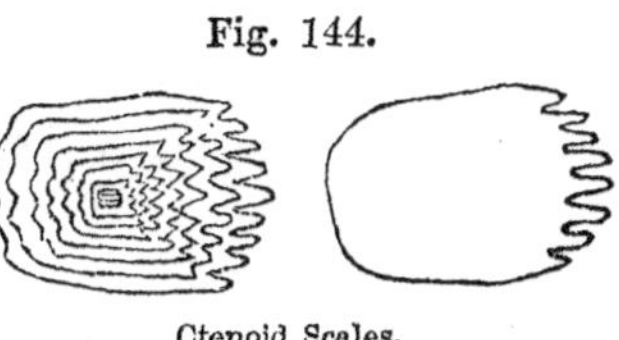

Ctenoid Scales.

262. The *reptiles*, like the fishes, shells and plants of this period, resemble those now living in warm insular climates; they include crocodiles, turtles, tortoises, and serpents similar to the boa-constrictor attaining sometimes the length of twenty feet. Several species of *birds* have been obtained from the Eocene resembling the pelican, buzzard, owl, woodcock, quail, &c.

263. The number and peculiar features of the *land quadrupeds*, found in the gypseous deposits of the Paris basin attracted the attention of Cuvier, whose accurate descriptions of extinct forms have given great importance to palæontology. These quadrupeds belonged principally to the order *pachydermata*, or thick skinned animals, represented by the elephant and horse of the present period. The Eocene quadrupeds which have attracted most attention belong to the genera *Palæotherium* and *Anoplotherium*. This period has been designated *the age of mammals*. (§ 152.)

264. The *palæotherium* (*palaios*, ancient; *therion*, wild beast,) resembled generally the tapir of the present day, having a short fleshy proboscis; in some of its zoological characters it was like the rhinoceros. Its feet were divided into three toes, instead of four, as in the tapir. The skeletons of the palæotheria are found so abundant and well preserved in the Tertiary rocks as to leave no doubt respecting their zoological characters. Twelve species have been found, varying in size from that of the rhinoceros to that of the hog. The largest figure in cut 145 represents this animal.

Fig. 145.

Eocene Quadrupeds.

The *Lophiodon* was a genus still more closely allied to the tapir; only imperfect fragments of its skeleton have been obtained, indicating, however, that some of its species attained the size of the rhinoceros.

265. The name of the *Anoplotherium* (*a*, without; *oplon*, weapon; *therion*, wild beast) defenceless animal, is

derived from its destitution of tusks or canine teeth longer than the incisors, in which respect it differs from all mammiferous animals except man; its feet terminated in *two* toes, and it had no proboscis. The genus embraces several groups, one species of which resembled the otter, but was much larger; its peculiarity was the great length of its tail, which surpassed that of the body; no animal of the present day except the kangaroo has so long and powerful a tail. A species of another group resembled the hare in dimensions and in the proportions of its limbs. But the most characteristic species is the Anoplotherium *Gracile*, called also *Xiphodon*, whose graceful and elegant form resembled that

Fig. 146.

Anoplotherium Gracile—(*Xiphodon.*)

of the gazelle; it was closely allied in zoological characters to the ruminating animals, (the deer tribe.) The true ruminants, however, as the camel, ox, sheep, goat and deer were absent at this period

Fifty-seven species of mammalia have been recognized in the Eocene, including a number of carnivorous quadrupeds, as the wolf, fox and raccoon. The opossum, bat and monkey are also found in the rocks of this period; the remains of the monkey, *macacus eocænus,* occurs in a bed of sand underlying the London clay, in England.

266. The fossil *insects* found in the rocks of this period present remarkable instances of the preservation of the most delicate objects; the nerves of the wings, the pubescence of the head, and some traces of the coloring are visible. The wings of the beetles are found in some cases extended beyond the wing-covers, indicating that the insect fell into water while flying. Seventy genera have been found at Aix, in Provence, including some species identical with those inhabiting that locality at the present time.

Fig. 147.

FOSSIL INSECTS.

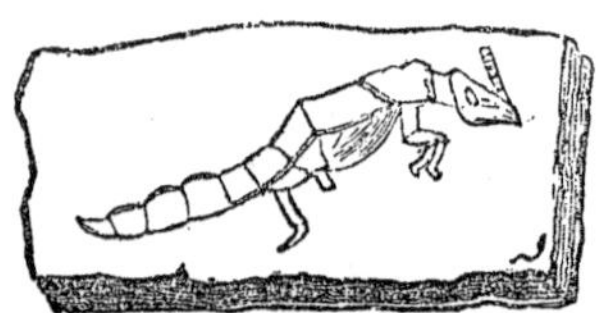

Lathrobium.

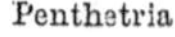

Penthetria.

Liparus.

267. The Eocene of North America is found in Virginia, Carolina and Georgia, and consists of beds of marl,

limestone, sands and clays; about six per cent. of its fossils are identical with the European Eocene fossils, but about one-fourth are very closely allied to them. The Eocene deposits of South America indicate a great difference in fossil contents on opposite sides of the Andes.

THE MEIOCENE PERIOD.

268. The *Meiocene*, called also the Middle Tertiary, is imperfectly represented on the east coast of England by the "Coralline Crag" formation, consisting of marine calcareous sands, thin limestones and marls; but in central and eastern Europe it is developed in strata of great thickness and extent. It occupies the great valleys of the Loire and Garonne in France; large portions of Switzerland; the valley of the Danube, including the Vienna basin, with the plains of Hungary and Poland; the valley of the Rhine and the western coast of Spain. The strata consist of quartzose sandstone, sometimes soft and incoherent, called in Switzerland the *Molasse*, shelly limestones and marls, with beds of lignite, and thick masses of corals. They are of both marine and fresh-water origin. The Meiocene beds have also been investigated in the southeastern States of the United States. Tertiary deposits on the flanks of the Sewalik hills in Northern India, and in Siam are referred to this period.

269. Extensive beds of lignite indicate a vigorous growth of *plants;* among them the palm is recognized, having grown in Central Europe contemporaneously with trees scarcely distinguishable from those still growing there. Among animal remains those of zoophytes were very abundant, making up entire masses of the coral crag, and belonging mostly to extinct species. Great numbers of shells,

marine and fluviatile are found; many of the latter belong to the genera *Limnea* and *Planorbis*, and are identical with those of the present day. The palæontology of this period combines the more ancient characters of the Eocene with the present botanical and zoological features of the same localities. The valley of the Rhine exhibits the most remarkable fossils of this period hitherto discovered. Numerous turtles and fishes presenting marked peculiarities of structure have been obtained there. Fig. 149 exhibits an interesting specimen three feet in length, from Oeningen near Lake Constance. This locality furnished also the celebrated relic, a gigantic, extinct, aquatic salamander, which Schewchzer in 1726 mistook for the skeleton of a man.

Fig. 148.

Fig. 149.

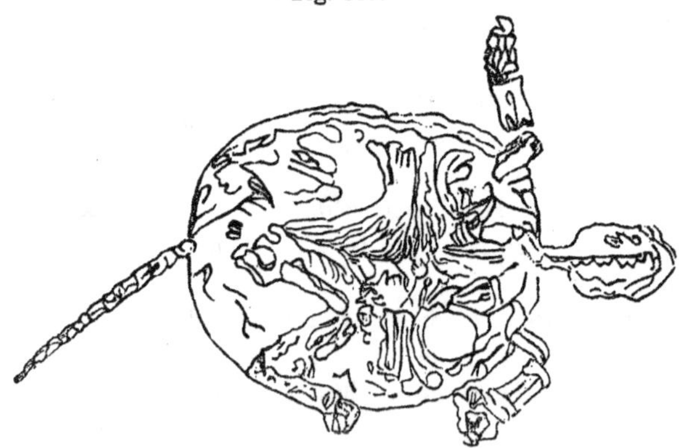

Fresh-water Tortoise.

270. The most remarkable quadruped of this period is the *Dinotherium*, (*deinos*, fearfully large; *therion*, beast,) whose remains are found in the valley of the Rhine near

Darmstadt, and in the valleys of the Jura Mountains. "Its length was nearly twenty feet; its body huge and barrel shaped, very much resembling that of the hippopotamus,

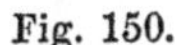
Fig. 150.

Dinotherium.

being little raised above the ground, although the huge columns which formed its legs are supposed to have been nearly ten feet in length. Its head, rarely, perhaps, brought entirely above the water, was like that of a large elephant, and it was provided with a short, but very muscular and powerful proboscis. A pair of large and long tusks were appended to this skull, and curve downward, as in the walrus. The tusks do not proceed from the upper jaw, whence they could be made to depend entirely upon the bones of the neck to support them, but are fixed in the lower jaw, and are planted, as it would seem, in this strange position at the greatest possible mechanical disadvantage. There can scarcely be a doubt that an animal provided with appendages so placed, was an inhabitant of the water; and the tusks, which are very large, were probably useful as pick-axes, enabling the monster to dig for succulent vegetable food by day, while, perhaps, at night they could be attached like anchors to the banks of the river or lake in which the animal habitually dwelt. It was the most gigantic of the herbivorous quadrupeds, and was associated with the palæotheres of the more ancient Tertiary period, and with the mastodons and elephants which lived on till a far more recent date."*

271. The Tertiary deposits upon the flanks of the sub-Himalayas or Sewalik hills in India abound with interesting fossils, including the monkey, elephant, dinotherium, mastodon, rhinoceros, hippopotamus, horse, giraffe, antelope, and many others resembling the existing animals of the same name. The most singular animal hitherto found in that locality is the *Sivatherium*, so called from the Indian god Siva, and *therion*, beast. Its size was about that of a rhinoceros. Its head was large and shaped like the elephant's, and was furnished with a small proboscis. It had two pairs of horns; one pair, like those of the ox in shape, were situated over the eyes, and the other pair palmated, like those of the elk, were placed on the back part of the head. The structure of the animal shows it to have been intermediate between the pachydermata and ruminantia.

The *Toxodon*, (*toxon*, a bow; *odous*, a tooth,) a large quadruped distinguished for the singular shape and arrangement of its gnawing teeth; and the *Macrauchenia* (*makros*, long; *auchen*, neck,) a pachydermatous animal whose neck was nearly as long as that of the giraffe, have been found in the Tertiary of South America.

THE PLEIOCENE PERIOD.

272. The only representative of the Pleiocene or Newer Tertiary in England is the *Red Crag*, consisting of ferruginous-colored sand and gravel, with few zoophytes, but abounding in marine fossil shells, many of which are iden-

* Ansted's Ancient World, p. 282.

tical with living species. On the Continent of Europe the extensive beds on both sides of the Apennine Mountains, called the Sub-Apennine deposits, are 2000 feet thick, consisting of calcareous marls and sands, and are extremely fossiliferous. Some of the deposits in Sicily, Greece and Asia Minor belong to this period. The Pleiocene strata cover also extensive regions in southern Russia, central and southern Asia, and America.

273. Remarkable beds of *lignite* or *brown coal,* belonging to this period, found in Germany, exhibit a great mass of vegetable matter. These lignites lie between beds of clay and sand, and consist of solid wood, blackened, but in many instances so slightly changed as to admit of being used as timber; they are similar to the trees now growing in their vicinity. A thin leafy bituminous lignite is called paper coal. The remains of fishes, frogs, insects and quadrupeds are found in these deposits.

THE PLEISTOCENE PERIOD.

274. The *Pleistocene* embraces the deposits of fossiliferous sands, marls and gravels, containing fossils which belong almost entirely to living species. They are extensive in the Southern parts of Europe, Asia and America. By some writers on Geology they are made to include the drift, while others apply the term to beds of clay, sand and marl deposited subsequent to the drift and previous to the Alluvial. Where the drift does not occur, as in the southern parts of Europe and North America, the Pleistocene beds present an uninterrupted series from the Pleiocene to the Alluvial.

275. The deposits which are spread over those vast plains in South America, called Pampas, are assigned to

the latter portion of the Tertiary period. They include a remarkable group of animals, belonging to the order Edentata, of which the sloth, the armadillo and the ant-eater are representatives in the fauna of the present day. The *Megatherium* (*mega*, great; *therion*, beast,) was a gigantic quadruped exhibiting some very striking resem-

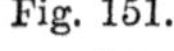

Fig. 151.

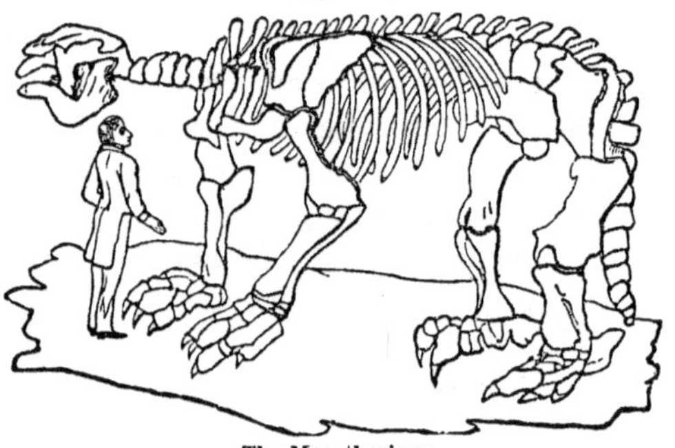

The Megatherium.

blances to the sloth. Its length is eighteen feet, its breadth across the loins six feet, while its height was nine feet. Its proportions were singularly massive, its pelvis and hind extremities being three times as large as those of the elephant. The fore legs were long, resembling in structure those of the sloth. The fore foot was a yard long, twelve inches broad, five-toed, and armed with long and powerful claws. This animal is taken as the type of the Megatheroid group.

276. The *Mylodon* (*mule*, a mill; *odous*, a tooth,) an allied genus found in the same locality, in some respects more closely resembles the sloth. Prof. Owen has given a full description of a perfect skeleton of this animal obtained from Buenos Ayres, and deduced its relations and habits. The length of the skeleton is eleven feet, inclu-

ding the tail. Its teeth show that it was a vegetable eater; it probably lived, as do the sloths of the present day, on leaves and buds of trees.

Fig. 152.

Mylodon Robustus.

277. The *Megalonyx* (*megale*, great; *onux*, claw,) resembled the mylodon in its size and proportions; it differed in its claws, and in greater freedom of motion in its fore limbs.

The *Scelidotherium* (*scelidos*, the thigh; *therion*, beast,) though allied to the megatherium, resembled the ant-eater and armadillo. In its hinder extremities and tail, the strength was greater in proportion than in any known animal, living or extinct; while the length of the animal was not greater than that of a Newfoundland dog, and its fore limbs no larger, its hind extremities were as ponderous as those of the hippopotamus.*

278. Associated with the megatheroid animals of this period was a colossal armadillo, called the *Glyptodon* (*glup·tos,* sculptured; *odous,* tooth,) from its teeth being sculptured laterally with wide, deep grooves. Like the armadillo

Fig. 153.

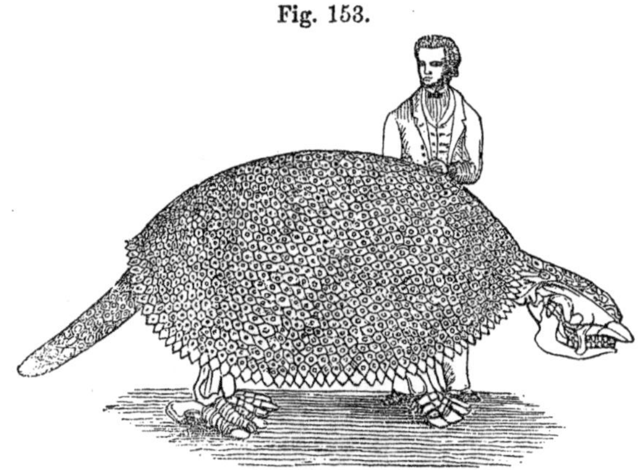

Glyptodon Clavipes.

it was covered with a crust or shell, composed of polygonal pieces accurately fitting each other, constituting a massive armour. The structure of its hind foot is very peculiar, presenting a frame work unparalleled in its adaptation to support a great weight, and at the same time allow such free motion in the fore limbs as the habits of the armadillo require. Several species of the genus glyptodon have been determined. The remains of the mastodon have been found in the same localities with the megatheroid animals.

* Ansted's Ancient World, p. 348

279. *Igneous* rocks are found abundantly in connexion with the Tertiary series; they are lavas of *extinct volcanoes*, and are intermediate in character between the traps of the secondary period and the products of active volcanoes. In composition they are principally trachytic, but sometimes scoriaceous, or tufaceous. One of the most remarkable localities of Tertiary volcanoes is an extensive plain in Auvergne, central France, which supports a series of seventy volcanic cones, varying from five hundred to a thousand feet in height, forming a range about twenty miles long and two broad. Many of these cones retain well defined craters several hundred feet deep, and their lava beds can be traced as easily as those of Vesuvius. "There is no spot," says Mr. Scrope, "even among the Phlegræan fields of Italy, which more strikingly displays the characters of volcanic desolation. Although the cones are partially covered with wood and herbage, yet the sides of many are still naked; and the interior of their broken craters, rugged, black and scorified, as well as the rocky floods of lava with which they have loaded the plain, have a freshness of aspect, such as the products of fire alone could have so long preserved, and offer a striking picture of the operation of this element in all its most terrible energy." These volcanic vents are of different ages; some of them are manifestly of comparatively recent origin.

280. Another interesting group of Tertiary volcanoes is upon the banks of the Rhine, where the graceful forms of the Siebengebirge or Seven Mountains, and the majestic Eifel with its crater covered with scoriæ, and its lava currents still visible, are conspicuous features in the picturesque scenery of the river. The basis rocks supporting the Tertiary formations are not here, as in Auvergne, granite,

but belong to the Silurian formation. Similar evidences of volcanic agency during the Tertiary period are found in Catalonia in Spain, Hungary, Asia Minor in the "burnt district," and in Palestine. The soil in the vicinity of these volcanoes, composed in part of their lavas, is exceedingly fertile.

CHAPTER IX.

ROCKS OF THE QUATERNARY PERIOD.

281. THE Quaternary series embraces the heterogeneous masses of the Drift and all subsequent deposits up to the present period. English geologists generally discard this division, and extend the Tertiary over this series.

THE DRIFT PERIOD.

282. The *Drift* includes accumulations of sand, gravel, clay, pebbles and boulders, or erratic blocks; the boulders are fragments of the hard crystalline rocks, usually water-worn, rounded, and scratched, or grooved. The term *Diluvium* has been applied to this deposit; drift indicates that the materials have been impelled by currents.

283. The drift is not universally *diffused*, but appears to be confined within 40° or 50° from the poles. In Asia it is rarely found lower than 60° north latitude; in Europe it descends to the southern parts of Poland and Prussia; and in North America it is found as low as 40°, and some of the north and south valleys extend it a little farther southward. Its southern limit in the United States is a line drawn from Long Island through central Pennsylvania to the Ohio, with occasional outliers in the valleys of the Delaware, Susquehanna and Mississippi. It is found also in the southern portions of South America, but not within the tropics. It occurs at *elevations* above the present sea

level, varying from three hundred to five thousand feet. The general *direction* of the drift in North America has been from north to south, and occasionally southeast; while in Europe it appears to have been dispersed in various directions from the Scandinavian Mountains. Boulders have been transported in some instances hundreds of miles; the largest of them have, however, usually been deposited within a few miles of the ledges from which they were torn. They diminish in size and number as the distance from their original position increases. Boulders of considerable size are frequently found in Northern Ohio, but are few and small in the central and southern parts of the State. They can be identified with the rocks of the parent ledges by their general constitution and by particular minerals contained in them.

284. Boulders vary greatly in *size*; some of them weigh thousands of tons. A conglomerate boulder at Fall River, President Hitchcock states, weighs ten million eight hundred thousand pounds; a granite boulder near Neufchatel weighs three million eight hundred and fifty thousand pounds; and the boulder from which the pedestal of the statue of Peter the Great was hewn, weighed one thousand five hundred tons. Some boulders are so poised upon hard surfaces, as to oscillate by the application of a moderate force to them, and are then called *rocking stones;* others are firmly posited in such a manner as to have given rise to the conjecture that they are artificial structures. An example of this occurs in the State of New York, and is represented in Fig. 154. A boulder of felspathic granite weighing about fifty tons, rests at the height of three or four feet above the ground, upon limestone pillars. They are sometimes poised on the summits of mountains; others

have been carried over mountain ridges, though they are usually found accumulated in larger numbers on the north sides of mountains.

Fig. 154.

Granite Boulder on Limestone Pillars.

285. The transport of the drift has produced very conspicuous effects upon the surface of the earth, scratching, grooving, and polishing the rocks. The scratches and grooves are parallel, resembling those produced by glaciers described in § 32. They vary from a fraction of an inch to more than a foot in depth and width. Sometimes two or more sets of striæ cross each other at a small angle. These striæ have been found on the White Mountains five thousand feet above the ocean level. Prof. Locke gives an example of these grooves on the limestone near Dayton, Ohio. "The quarry has been stripped of soil more or less over ten acres, and the upper layer of stone is in most

places completely ground down to a plane, as perfectly as it could have been done by a stone cutter, by rubbing one slab on another with sand between them. In many places, in addition to the planishing, grooves and scratches in parallel straight lines, evidently formed by the progress of some heavy mass, propelled by a regular and uniform motion, are distinctly visible. The grooves, are, in width, from lines scarcely visible, to those three-fourths of an inch wide, and from one-fortieth to one-eighth of an inch deep, traversing the rock in a direction south 26° east, in lines exactly straight and parallel."*

It is probable that a very large portion of the earth's surface was affected in this way, as far as the drift extended, since the removal of clay and other substances which cover the rocks discovers the striæ; rocks that suffer disintegration by atmospheric agency, as limestones, which are partially dissolved by water holding carbonic acid in solution, do not retain their grooves. The direction of the striæ coincides with that of the dispersed drift, and is often modified by the features of the surface; as when it is diverted from the general direction into a valley. Some sets of striæ appear to have been nearly obliterated by others passing over them.

286. The northern and northwestern sides of the ledges of rocks are more worn by the drift agency, and the hills are elongated in this direction, corresponding in appearance with those denominated by European writers *roches moutonnces.* President Hitchcock has observed very numerous angular fragments of rocks ranged in long narrow lines, extending from the ledges in the same direction with the drift, and overlying that deposit, which he denominates

* Ohio Geological Survey, p. 230.

streams of stones. The same geologist adduces as instances illustrative of the prodigious violence of the drift agency, the fracture and overturning of perpendicular strata of slate rocks near the summits of hills. As the materials of the drift are generally supposed to have been transported by the agency of ice, these deposits are called *glacial beds;* they have very few, if any fossils.

287. Overlying the boulder formation occur beds of blue and yellow clay, sand and marl; these are most abundant in lakes, ponds, and river valleys. They sometimes appear to be caused by a new arrangement or assorting of the materials of the drift, producing an inter-stratification of sands, clays, and gravel, and are called *altered drift.* This appears to have been accomplished beneath the ocean which prevailed over the drift region.

Fig. 155.

Parallel Roads and River Terraces.

288. The origin of *marine terraces* has been assigned to the agency of the ocean exerted at this period. Fig. 155 presents a view in the valley of Glen Roy in Scotland, in which two parallel shelves or terraces, level and contin-

uous through the whole glen, are seen. They vary in width from ten to sixty feet, and are covered with boulders. The highest one is one thousand two hundred and fifty feet above the present level of the ocean, and the other is two hundred feet lower. These terraces are stratified deposits; they are ascribed to the action of water standing at that level, either a lake or the ocean. Similar shelves at these elevations are found in other valleys of Scotland, in Sweden, and in North America. *River terraces* present similar phenomena; they occur in valleys of mountainous districts, where the river flowing over the drift in which it cuts its channel, removes the materials to lower levels. The sudden removal of obstacles gives origin to a new terrace. These are represented in Fig. 155. Ancient lakes have also been reduced by successive stages, and formed broad level terraces.

289. Various beds of sand and gravel scattered over the valleys and plains are called *ossiferous*, because they contain bones of the elephant, hippopotamus, bear, deer, horse and other animals which do not now inhabit the regions where their remains occur; the bones of the elephant and rhinoceros are found in England and in Siberia, where they have not been known to exist within the historic period. These bones are partially petrified by the salts of lime and iron, are harder and heavier than recent bones, but still preserve their bony structure. Many of these ossiferous deposits are local, having been produced by the action of rivers, and the filling up of lakes, but some of them appear to be due to more extensive agencies.

290. Numerous *caverns* have been found in Europe, America and Australia, containing deposits of loam, river silt and small boulders. These materials were probably

introduced during different periods; but the animal remains included in them indicate the drift period as the one during which the largest portion of the deposits accumulated. The bones, which are perfectly preserved in many instances in calcareous incrustations, are chiefly those of races of bears and hyænas which inhabited the caves, together with the remnants of their prey, and occasional fragments of the elephant and rhinoceros. The remains of man, and of animals still living in the vicinity are sometimes found in them. The most remarkable caverns, on account of their organic contents, are Kirkdale Cave near York in England, and the Cave of Gailenreuth in Germany.

291. The *Kirkdale Cave,* which Dr. Buckland has very accurately described in his "Reliquiæ Diluvianæ," is situated about twenty-five miles northeast of York, above the northern edge of the great vale of Pickering, and thirty feet above its waters. Its floor is level and nearly conformable to the plane of stratification of the coralline oolite in which it occurs. In some parts the cave is three or four feet high, and roofed, as well as floored, by the level beds of this rock; in other parts its height was augmented by open fissures, which communicate through the roof, and allow a man to stand erect. The breadth varies from four or five feet to a mere passage; at the outlet or mouth against the valley was a wide expansion or ante-chamber, in which a large proportion of the greater bones, as of the ox, rhinoceros, &c., were found. This mouth was choked with stones, bones and earth, so that the cave was discovered by opening upon its side in a stone quarry. On entering the cave, the roof and sides were found incrusted with stalactites, and a general sheet of stalagmite, rising irregularly into bosses, lay beneath the feet. This being broken

through, yellowish mud was found about a foot in thickness, fine and loamy toward the opening, coarser and more sandy in the interior. In this loam chiefly, at all depths, from the surface down to the rock, in the midst of the stalagmitic upper crust, and as Dr. Buckland expresses it, "sticking through it like the legs of pigeons through a pie-crust," lay multitudes of bones of the following animals:

Carnivora—hyæna, tiger, bear, wolf, fox, weasel.

Pachydermata—elephant, rhinoceros, hippopotamus, horse.

Ruminantia—ox, three species of stag.

Rodentia—hare, rabbit, water-rat, mouse.

Birds—raven, pigeon, lark, duck, snipe.

The hyæna's bones and teeth were very numerous—probably two or three hundred individuals had left their bodies in this cave; remains of the ox were very abundant; the elephants' teeth were mostly of very young animals; teeth of the hippopotamus and rhinoceros were scarce; those of water rats very abundant. The bones were almost all broken by simple fracture, but in such a manner as to indicate the action of hyænas' teeth, and to resemble the appearance of recent bones broken and gnawed by the living Cape hyæna. They were distributed as in a "dog-kennel," having clearly been much disturbed, so that elephants, oxen, deer, water-rats, &c., were indiscriminately mixed; and large bones were found in the largest part of the cavern. The teeth of hyænas were found in the jaws, of every age, from the milk tooth of the young animal to the old grinders worn to the stump; some of the bones were polished in a peculiar manner, as if by the trampling of animals."

292. The most remarkable ossiferous cavern of Germany is that of Gailenreuth, which lies upon the left bank of the Wiesent. The entrance, which is about seven feet high, is in the face of a perpendicular rock, and leads to a series of chambers from fifteen to twenty feet high and several hundred feet in extent, terminated by a deep chasm, which, however, has not escaped the ravages of visitors. This cavern is perfectly dark, and the icicles, or pillars of stalactite, reflected by the torches which it is necessary to use, present a highly picturesque and striking effect. The floor is literally paved with bones and fossil teeth; and the pillars of stalactite also contain osseous remains. Loose animal earth, abounding in bones, forms in some parts a layer ten feet in thickness. A graphic description of this cave was published by M. Esper, more than sixty years ago; at that period some of the innermost recesses contained wagon loads of bones and teeth; some imbedded in the rock, and others in the loose earth. The bones in general are scattered and broken but not rolled; they are lighter and less solid than recent bones, and are often incrusted with stalactite. Cuvier, who enjoyed the opportunity of examining a very large collection of bones from Gailenreuth, was enabled to determine that at least three-fourths of the osseous contents of the caverns belonged to some species of bear; and the remaining portion to hyænas, tigers, wolves, foxes, gluttons, weasels and other small carnivora. By the bones which were referable to the bear, he established three extinct species of that genus, the largest of which is called the *Ursus spelæus*. The hyæna was allied to the spotted hyæna of the Cape, but differed in the form of its teeth and head. Bones of the elephant and rhinoceros are also said to have been discovered, together with those of exist-

ing animals, and fragments of sepulchral urns of high antiquity."*

293. The phenomena of ossiferous caverns lead to the conclusion that they were the dens of ravenous animals; that the carcasses of large animals were drifted into them; and that men have used them as places of abode or sepulture. Fragments of bones, mingled with clay, pebbles, shells, etc., cemented together with the carbonates of lime and iron, are frequently found filling fissures in the rocks. Such accumulations are called *osseous breccia;* they are very abundant in the vicinity of the Mediterranean sea; the rock of Gibraltar yields fine specimens. The bones of the breccia are referable to both extinct and recent species. The bone breccia of Australia has the same ochreous color as that in Europe has; its bones are all referable to marsupial animals, as the *kangaroo, wombat, dasyurus,* &c.

294. The fossils of this period are very numerous and various; shells, both marine and fresh-water, are found in great quantities beneath beds of gravel and boulders, and especially in beds of marl under the muck—decaying vegetable matter—of ponds and swamps. The long clam, *Mya arenaria,* and the common oyster, *Ostrea borealis,* marine shellfish, are found in the deposits of this period far inland, and a very large majority of the shells belong to species inhabiting the ocean of the present day. The immense accumulations of these shells, constituting layers many feet deep, indicate the lapse of a long period of time. *Infusoria* abounded at this period; the silicious marl beneath peat swamps is almost entirely made up of these fossil skeletons.

295. At this period the most gigantic of the existing

* Mantell's Wonders of Geology, p. 169.

groups of herbivorous quadrupeds were represented by allied species, which had a very wide range of diffusion, and some of which had a special organization to enable them to encounter a severity of climate, such as similar animals are not at the present day designed to endure, the elephants, for example, having been covered with hair. The animals which formerly characterized the plains of Siberia and the high latitudes of the North American continent are closely allied to the existing fauna of Northern Europe.

296. The *Mastodon* (*mastos*, nipple; *odous*, tooth,) is a genus quite distinct from the elephant and derives its name from the crown of the molar tooth presenting conical tubercles covered with enamel. It was a somewhat larger animal than the elephant, with a body longer in proportion. It was very widely diffused, its remains having been found in Asia, Europe, and America, from the equator to 66° of north latitude; it has also an extended range in time, connecting the Meiocene with the Pleistocene deposits, and continuing down nearly or quite to the epoch of man, concomitant in the latter periods of its existence with many species of animals which still survive. The temperate zone of the North American continent appears to be the locality in which it flourished; it is there five times as numerous as the elephant. Its remains abound in the marshes whose waters are saline, called *Licks;* the skeletons have been found erect with the head thrown upward, as if the animal

Fig. 156.

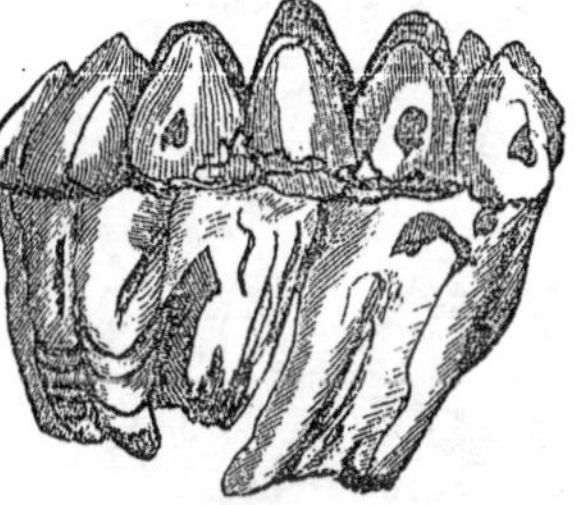

Tooth of the Mastodon.

had sunk in the mire; the stomach with its contents of bruised twigs and leaves has been found, confirming the conclusion which the structure of its teeth had suggested, that it was an herbivorous animal feeding upon tough coarse vegetables, as the branches of trees. It is estimated that from the Big Bone Lick in Kentucky, the bones of one hundred mastodons, twenty elephants, two oxen, two deer and one megalonyx have been extracted. Several very fine skeletons of mastodons are now in the cabinets of Europe and America. Fig. 157 presents a likeness of the most perfect skeleton yet exhumed. It was obtained at New-

Fig. 157.

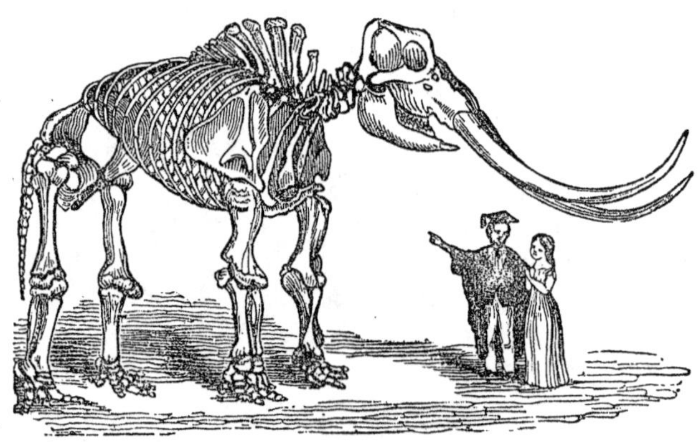

Mastodon found in Newburg, N. Y.

burg, N. Y., in 1845, and presents nearly every bone perfectly preserved; it is about twelve feet high, its tusks are fourteen feet long, and it weighs two thousand pounds.

297. The *Mammoth* (from the Arabic *behemoth*, signifying elephant,) was a companion of the mastodon from the Eocene period to the close of the Pleistocene, at the

close of which, like the latter, it became extinct. It differed from the mastodon particularly in the disposition of the enamel of its teeth in vertical plates or layers alternating with softer bone. It differed from the two existing species of elephant—the African and Asiatic—but was more nearly related to the latter. In the crown of the tooth of the African species the enamel is arranged in lozenge-shaped figures, as represented at *a* Fig. 158; in the Asiatic species the enamel is in narrow transverse bands, as at *b*; while the enamel in the tooth of the fossil species is similarly arranged in broader bands, as at *c*. The remains of the fossil elephant are very numerous, occurring wherever the mastodon is found, but relatively much more abundant on the Eastern than on the Western Continent. Bones of more than five hundred individuals are supposed to have been found on the coasts of Norfolk and Suffolk in England, and they are very abundant in the shoals of the German Ocean. They have also been found in Ohio, Vermont, and other localities in the United States. But the most singular localities in which these fossils have been found are the frozen gravels and clays at the mouths of rivers and along the shores of the Polar seas, in latitudes in which the existing species of elephant can not live. The skeleton of the fossil elephant found in Siberia, described in § 124, closely resembles that of the Asiatic species, but its tusks are larger and more curved backward. Its tusks weighed three hundred

Fig. 158.

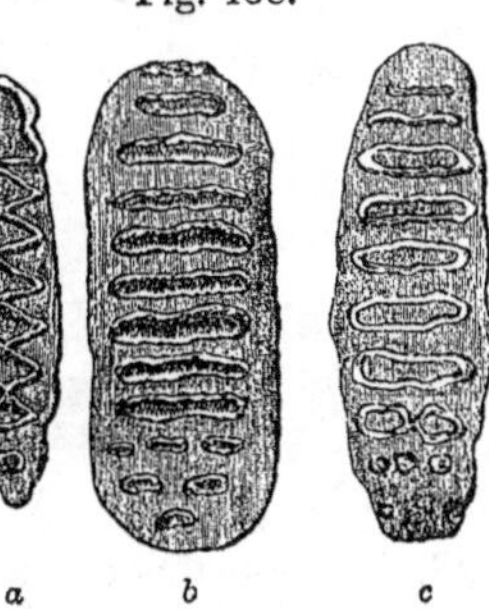

and sixty pounds, and the head with the tusks four hundred and fourteen pounds. A large portion of the ivory of commerce has been derived from the fossils in Siberia.

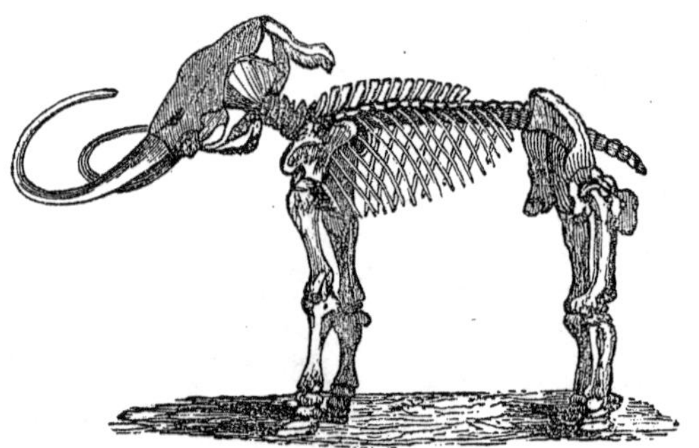

Elephas Primigenius.

298. The remains of other extinct animals are found upon the shores of the Arctic ocean, one of which bears resemblance to the rhinoceros but is quite distinct from any known species of that genus. Von Wrangel states that on an island in latitude 75° and longitude 140°, in the Polar sea, the hills in the interior were found to contain the skulls and bones of horses, buffaloes, oxen and sheep in such abundance that these animals must formerly have lived there in large herds. The best bones as well as the greatest number are found at a certain depth below the surface, usually in clay hills, more rarely in black earth. The more solid the clay, the better are the bones preserved, and experience has shown that more are found in elevations situated near higher hills, than along the low coast, or on the flat tundra (moss levels.)*

299. The bones of ruminating animals nearly allied to

* Ansted's Geology, II, p. 151.

the common species of the ox and deer tribes are very abundant in the deposits of this period; the dimensions of their bones indicate great size, but otherwise they differed slightly from existing species. A remarkable gigantic animal, called the Irish Elk, is found abundant in the east of Ireland and in the Isle of Man. It is found in beds of marl beneath peat-bogs which are the sites of ancient lakes; they occur also in the marls of France, Germany and Italy. The species is remarkable for the dimensions of its antlers, which were palmated. The average weight of the skull and antlers is seventy-five pounds. The

Fig. 159.

Megaceros Hibernicus.

skeletons are ten feet high, and the antlers spread from ten to fourteen feet. This animal continued to inhabit the earth up to a late period, and by some geologists is supposed to have been in existence after the introduction of the human race. Contemporary with this and other members of the deer tribe were several species of the genus *Bos*, of which one is supposed to be the great auroch or wild bison, still living in the forests of eastern Europe.

300. Besides the land mammalia, relics of marine tribes of the class are found in the Tertiary, in the Pleiocene, and perhaps in the Meiocene, but more abundantly in the Pleistocene, in beds of drift in valleys traversed by rivers. The *Cetacea* are not fishes; they breathe not by means of gills, but by lungs, are viviparous and suckle their young.

Dr. Harlan described the skeleton of an animal of enormous size found in the Tertiary in Alabama, under the name of *Basilosaurus* supposing it to be a reptile; but Professor Owen has shown that it is a Cetacean, allied to the Dugong and Cacholot, and has assigned to it the name *Zeuglodon.*

The skeleton of a whale seventy-two feet long has been discovered in a clay bed in Scotland, twenty feet above the present high tide level.

A portion of the skeleton of a whale which was sixty or seventy feet long, has been extracted from a cliff in Brighton, England, associated with the remains of the mammoth.

The skeleton of a whale thirteen feet long has been recently discovered by a railroad excavation in the blue clay of the valley of Lake Champlain, Vermont.

301. The skeletons of several species of *birds* equal in size to those which are supposed to have made the footprints upon the New Red Sandstone, have been found in the recent deposits in New Zealand. They appear to belong to a group of which the Apteryx is a living representative, which are more ponderous in proportion than the ostrich, and destitute of wings. They vary greatly in size,

some having attained the height of nearly twelve feet, while others are quite small. The structure of the Dinornis shows that it was, like the Apteryx, well fitted for rapid running. These birds appear to have been continued into the historical period; some of them have very recently become extinct; their bones are found in the state denominated sub-fossil.

Fig. 160.

The Dinornis.

302. The *climate* of the period immediately succeeding the Drift appears to have been essentially the same as it now is. The deposits of arctic shells in the altered drift which were thought to indicate a low temperature, are quite limited and may be accounted for by the influence of polar currents; many of them are identical with those of species now living exposed to such currents on the present sea coasts. The existence of elephants, lions, tigers, hyænas, and similar animals found now only in tropical regions, indicates a mild climate in Great Britain and other parts of Europe where their remains are found in abundance. The elephants found in Siberia were clad with wool and hair, showing that they lived in a colder climate than their congeners are fitted for at the present day; but the climate must have been milder than it is in those regions at present,

to have furnished sufficient food for such large animals as the elephant and rhinoceros, and the supposition that it was, appears to be confirmed by the fact that large birch trees are found embedded in the sandy cliffs, beyond the seventy-fifth degree of north latitude in sufficient quantities to be used by the inhabitants as fuel, while only stunted shrubs of the same genus grow at the present day beyond the seventieth degree of latitude.

THE ALLUVIUM.

303. The *Alluvium* embraces the recent and progressive formations, consisting of sands, gravels, clay, marls, vegetable and animal matter, which the rivers, lakes, seas, shell-beds and coral-reefs are constantly accumulating. These deposits are not distinguished from those which immediately precede them by any marked difference of characters, but through gradual, successive steps, they bring the history of the physical changes to which the globe has been subjected to the present time. They may be classified in accordance with their positions—as *marine*, *fresh-water*, and *terrestrial*. The period during which they have occurred is designated the *historic period*—the *age* or *reign of man*.

304. The phenomena of *raised beaches* indicate recent extensive changes of level in the ocean; such beaches, with their strewn sand, gravel, pebbles, shells, etc., occur in some instances hundreds of feet above the present ocean level. Some of the elevations have been sudden, and were witnessed by living observers, as on the coast of Chili in 1822, and the Ullah Bund, (§§ 67, 68;) but others are more ancient, and were probably gradual. Along the coasts of the Mediterranean, ancient beaches covered with the shells of species of shell-fish living at the present time in the ad-

joining sea have been elevated fifty feet, while the coasts of Sweden and Norway have been raised two hundred feet above the present level of the Baltic. Similarly located beaches or terraces are found near the ocean, generally conforming to its present boundaries, in all parts of the world.

Phenomena of a converse kind are presented by *submarine forests*, which are beds of vegetable substances with the roots of the trees in the situations in which they originally grew—now depressed several feet below the lowest tide; as the trees belong to existing species, the depression is shown to be a geologically recent occurrence.

305. Masses of sand, gravel, pebbles and clay, termed *marine silt*, have accumulated in many positions on the present shores of the ocean, through the influence of waves, tides and currents. The Isthmus of Suez is said to have gained thirteen miles in width within four thousand years; the sites of the ancient sea ports Tyre and Sidon are now several miles inland; and hundreds of square miles of dry land have been formed by the existing seas, in numerous localities, (§ 44.) Of the recent deposits made upon the bed of the ocean but little is known; soundings, however, show that sand, mud, shells, corals, and vegetable substances have been deposited by sub-marine currents on a scale which rivals some of the ancient strata. The Yellow Sea and the German Ocean have been shoaling at a very sensible rate within the period of human observation; the former, it is estimated, converts a square mile into solid land in seventy days, or more than five hundred square miles in a century.

306. *Estuary deposits* are peculiarly complex, since they are composed of marine and fluviatile silts, and embrace organic remains derived from the ocean, fresh-water,

and the land. The delta of the Niger may serve as an example: its base extends three hundred miles along the coast of the ocean, and consists of a beach of sea sand, with shells, corals and other marine remains; during the dry season, and at a low stage of the water in the river, these marine deposits extend farther inland. The delta extends up the river one hundred and seventy miles, consisting of vast expanses of low lands, swamps, and mud islands, separated by branches of the river and stagnant pools. A rank growth of marsh plants covers much of its surface; shell-fish and amphibious animals abound, and contribute their remains to its accumulating strata. Like the delta of the Nile it is annually inundated, and deposits of mud and sand are made over the whole surface, mingled with the remains of the elephant, hippopotamus, rhinoceros and other tropical animals, together with a variety of corresponding plants. When this delta shall have been elevated beyond the reach of the waters, and become consolidated, its strata of confused and alternating marine, fresh-water, and terrestrial characters, will present appearances similar to those of the Tertiary and Wealden formations.

Deltas are the most extensive of the accessible alluvial deposits; they furnish a connecting link between formations now in progress and those of former geological eras.

307. *Lacustrine deposits* include the mud, sand, marl and organic matter which have accumulated in fresh-water lakes of the present period; these materials form successive regular layers upon the bottoms of the lakes, indicating quiet, gentle deposition. Lakes have thus been filled up and formed flat alluvial plains; some geologists ascribe this origin to the *prairies, pampas,* and *steppes.* Some lakes have been partially or wholly *drained* by a more rapid pas-

sage of their waters through the outlets; when this has been effected by distinct stages, successive terraces are formed; this is supposed to have been the origin of the *lake ridges* —*ridge roads*—they having constituted the ancient shores of the lakes. The circumstances of deposition of the fresh-water beds of the Tertiary, Wealden, and Coal series appear to have been similar to those of lacustrine accumulations of the present era. Rivers have sometimes deposited their silt upon their beds, and more frequently during freshets upon the adjacent valleys; such deposits are more heterogeneous and irregular than those made in lakes.

308. Of the various mineral substances chemically precipitated from water, *marl* is the most abundant; it is the carbonate of lime held in solution and in mechanical suspension. It occurs in various degrees of purity; when densely aggregated and sometimes crystalline in texture, it is called *rock marl;* when cementing together a mass of shells, *shell marl*; and when largely mixed with clay, *clay marl.* Marls are found most frequently in the deposits of lakes, ponds and swamps in limestone districts, in which calcareous springs abound. Immense quantities of this material are also conveyed by rivers into the ocean. "A hard stratum of *travertin,*" says Mr. Lyell, "about a foot in thickness, is obtained from the waters of San Filippo in four months; and as the springs are powerful, and almost uniform in the quantity given out, we are at no loss to comprehend the magnitude of the mass which descends the hill, which is a mile and a quarter in length, and the third of a mile in breadth, in some places attaining a thickness of two hundred and fifty feet. To what length it might have reached it is impossible to conjecture, as it is cut off by a stream which carries the remainder of the calcareous matter

to the sea." Waters holding silica in solution—hot springs—have deposited *silicious sinter;* the hot springs of St. Michael have encrusted surrounding objects, and deposited layers of sinter several inches thick. *Alumina* is also sometimes similarly precipitated; a mixture of precipitated alumina, silica and the oxide of iron has been obtained abundantly from Tripoli for use as a polishing powder but much of the *tripoli* of commerce consists of the silicious shields of animalcules. The oxides of iron and manganese, gypsum, common salt, petroleum or asphaltum, etc., have formed extensive deposits; they furnish many analogies by which the phenomena of similar beds in the older strata are illustrated.

309. The *plants* embedded in the alluvium belong to existing species, and occur as *subterranean forests, peat-mosses,* and *drift-wood.* Subterranean forests occur in depressed valleys, and low alluvial plains, where trees standing erect or overturned in the situations in which they grew have been invested with mud and sand; the wood of these trees, though discolored, has been employed in the construction of houses in England. Rivers, ocean-currents, and tides have formed extensive accumulations of drift-wood in estuaries and deltas, and along the sea coast, particularly in sheltered bays. An instance of alluvial vegetable accumulation has been discovered by a section of a canal in Scotland. "At a depth of twelve feet from the surface of the fine alluvial sediment," says Professor Phillips, "a quantity of hazel bushes, roots, and nuts, with some mosses, fresh-water shells, and bones of the stag were met with. In some parts of the sediments an English coin was found, and oars of a boat were dug up. Where a little water entered this peaty and shelly deposit from the adja-

cent limestone, it produced in the wood a singular petrifaction; for the external bark and wood were converted into carbonate of lime, in which the vegetable structure was perfectly preserved. In like manner some of the nuts were altered; the shells and the membranes lining it were unchanged; but the kernel was converted into the carbonate of lime, not crystallized, but retaining the peculiar texture of the recent fruit. In this particular case no reasonable doubt can exist that the peaty deposit, full of land-mosses, hazel-bushes, and fresh-water-shells, was *water-moved,* and covered up by fine sediments from the river and the tide."

310. But the most extensive vegetable deposits of the alluvium are *peat-mosses,* which cover hundreds of square miles, and are sometimes forty feet thick. *Peat* consists of mosses, especially the sphagnum palustre, rushes and other aquatic plants, together with the trunks, branches, and leaves of trees. Peat swamps occupy the sites of ancient lakes and low woody districts, in which obstructions to the drainage have caused swampy morasses, destroying the forest trees and favoring the growth of aquatic mosses. The sites of many of the aboriginal forests of Europe are now covered by mosses and fens. Fallen trees by obstructing the drainage of a district have produced peat-bogs; the prostration of a forest by a tornado about the middle of the seventeenth century produced a peat-moss, which at the beginning of the eighteenth century yielded peat for fuel. Peat swamps possess eminent antiseptic power; the bodies of men and other animals buried in them have been preserved for centuries. The most ancient peat-mosses belong to the alluvial period; this is known by their conforming to the present configuration of the land, and by their containing the remains of vegetable and animal bodies belonging

exclusively to existing species. The changes which occur in peat-beds illustrate the formation of coal; a true bituminous coal has been found in a peat-bog in the State of Maine, several feet below the surface amidst the remains of logs of wood, (§ 144.)

311. The remains of *animals* belonging to the species still living or very recently extinct, are characteristic of the Alluvium; among these remains the exuviæ of shell-fish and coral zoophytes are, as they were in the older formations, most abundant. Immense banks of dead *shells* have been drifted together by tides and currents; some of

Fig. 161.

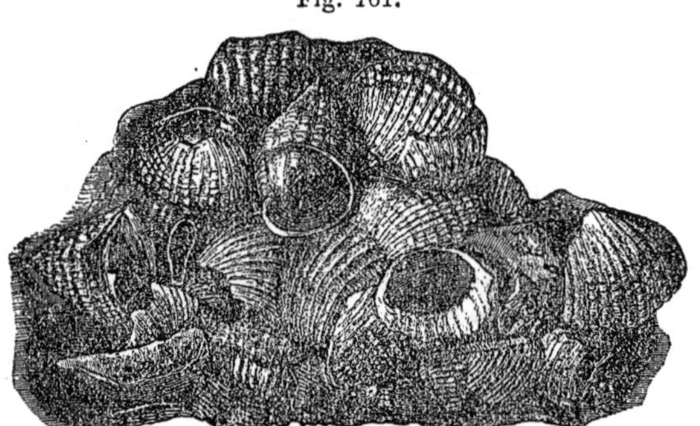

Shell Limestone from the Mouth of the Thames.

them remain loose and are worn by the waves, while others are cemented together by the carbonate of lime into a shelly limestone or sandstone. Fig. 161 represents a portion of a block taken from a bank of consolidated shells, in the progress of formation in the English channel near the mouth of the Thames; it is sufficiently firm to admit of being cut and polished; the bank consists mostly of one species, having lived gregariously, as do oysters and mussels. *Coral* reefs with their accumulated debris constitute thousands of square miles of the surface of the earth, (§ 76,) and form an aggregate of calcareous matter equal to the limestones of the older formations. The remains of fishes, reptiles, birds and mammalia, which are included in alluvial deposits, though comparatively few in number, are highly indicative of the circumstances in which they lived.

312. Some genera and species of animals have become extinct during the alluvial period; this may have been the case with the great Irish Elk, the Mastodon, the Dinornis and others; but there is one example of extinction which has occurred within the period of authentic history. The *Dodo*, a bird of the gallinaceous tribe larger than a turkey, abounded in Mauritius and adjacent islands, when first colonized by the Dutch, more than two centuries ago. It is now entirely extinct. The stuffed skins formerly in European Cabinets have decayed, and the only relics of them are a few fragments of the harder parts, as the head and feet, in the British Museums. The bones of the Dodo have been found fossil in a tufaceous deposit in the Isle of France. The *Apteryx*, so called because destitute of wings, is a rare bird living in New Zealand. It differs from most birds in many particulars of its organization. It appears to be almost extirpated, and several of its congeners are thought to have recently become extinct

Fig. 162.

The Dodo.

313. The remains of *man*, are found only in the newest deposits, nor is there any reason for supposing that he existed at an earlier period, since if he had, some vestiges of his existence would have been perpetuated. Cuvier observes that the bones of men in ancient battle fields were as well preserved as were those of the horses buried with them; and much frailer substances than human skeletons have been fossilized in very ancient rocks. The skeletons of men, as well as various works of art, have been found in the alluvium. The discovery of human skeletons in limestone has been referred to in § 91; they were found on the northeast coast of Guadaloupe in a bed of rapidly accumulating limestone, consisting of fragments of corals and shells, with sand cemented together by the carbonate of lime. The shells and corals belong to existing species, and the pieces of pottery and implements found with the skeletons identify the period of their deposition as historically recent.

Fig. 163.

The Apteryx.

Fig. 164.

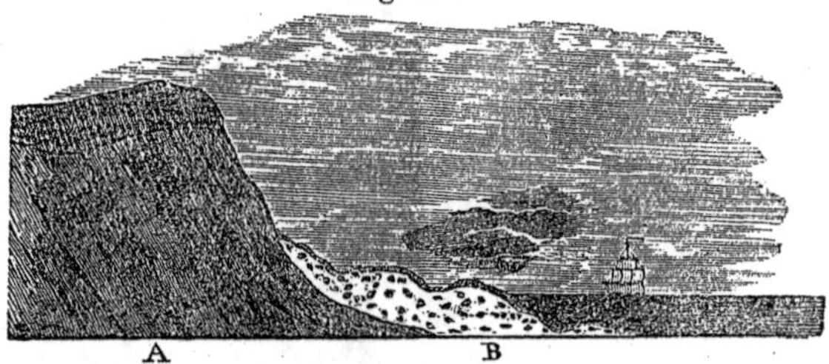

Cliff at Gaudaloupe.

A, the ancient rocks; B, alluvial limestone containing human skeletons.

The conjecture that man's remains will be found in the older rocks of Asia where the race was first introduced, is shown to be groundless by the fact that the organic contents of Asiatic rocks are known to correspond with those of the rocks of the same age in other parts of the world.

314. The superficial stratum of the earth's surface in which plants grow is denominated *soil;* it consists of minute fragments of rock—gravel, sand, clay, together with portions of decomposed vegetable and animal matter. It varies in thickness, never exceeding however a few feet. The fragments of which it consists, have either resulted from disintegration of the subjacent rocks or have been transported from the localities in which they were broken up. The mineral mass lying immediately beneath, and destitute of vegetable and animal substances, is called *subsoil.* Soils are designated according to their predominant mineral ingredients, as *sandy, gravelly, loamy, clayey or calcareous.* The soil not only furnishes subsistence to vegetable and animal bodies but protects the surface of the earth from rapid wearing; an unprotected surface is wasted by every shower, while even light sands are secured from abrasion by vegetables growing upon them.

315. Atmospheric and aqueous agencies exerted during the alluvial period, have produced the most recent modifications of the earth's features; the forms of mountains and valleys have been losing their sharp angles, and become undulating. *Caves,* which were originally fissures in limestone rocks, have been greatly extended by the eroding action of water. The great Kentucky Cave has been traced ten miles in one direction, without finding any termination, it has also very extensive lateral branches. A similar action

in gorges and valleys occasionally presents the phenomena of *natural bridges*, of which a remarkable example is found in Virginia. It consists of a magnificent arch of limestone spanning the Cedar creek. Its height above the stream is

Fig. 165.

Natural Bridge, Rockbridge County, Va.

two hundred and fifteen feet; its length ninety-three feet; its width eighty feet, and its thickness fifty-five feet.

316. The products of *igneous* agency during the allu

vial period are numerous and extensive beds of lava, volcanic ashes, scoriæ, sand, sulphur, etc. New volcanic craters have arisen, as Jorullo, in Mexico; while others have ceased to erupt. Earthquakes have produced great disturbances of the earth's crust, and various portions of the surface have been elevated and depressed. Many mountain

Fig. 166.

Chimborazo.

peaks show, by their forms, and by the lava upon their sides, that they are volcanic vents, though they have been long dormant, and can not be with certainty assigned to the alluvial, or period of active volcanoes.

CHAPTER X

THEORETICAL GEOLOGY.

317. HITHERTO the facts respecting the structure of the earth's crust, the nature of the materials and the order of their arrangement, have been detailed, constituting Descriptive Geology; but Geology when perfected embraces an enunciation of the *laws,* in accordance with which the successive changes the earth has undergone, and its present condition have been produced. This department of the science is termed *Theoretical,* or *Physical Geology.* Sir John Herschel observes, "The first thing that a philosophical mind considers, when any new phenomenon presents itself, is its *explanation,* or reference to an immediate producing cause. If that can not be ascertained the next thing is to generalize the phenomenon, and include it, with others analogous to it, in the expression of some law." *

A *theory* is a philosophical explanation of phenomena, deduced from principles which have been established by independent evidence; while an *hypothesis* rests solely on the satisfactory explanation of the phenomena, which it furnishes. These terms are, however, frequently used synonymously, and some modes of explanation deemed theories may by the progress of science be shown to be untenable by hypothesis.

* Herschel's Discourse on the Study of Natural Philosophy.

318. Geology is an inductive science based upon the observation of facts, and if the theories adduced to account for the phenomena are shown to be untrue or insufficient, the facts still remain unimpeached. Geological theories relate principally to the modes of action of the two great agencies *aqueous* and *igneous*. It has before been stated (§ 13,) that geologists are divided in opinion respecting the intensity with which these agents have operated in different periods; and while they agree in attributing the stratified rocks to deposition from water, in particular modes, the phenomena of the *Drift* have given rise to several theories. Diverse modes of explanation have also been proposed to account for the phenomena of volcanoes, and the elevation of mountains and continents. All correct reasoning in natural science is based upon the uniformity of nature's laws; a conviction of this uniformity is unceasingly impressed upon the mind by experience and observation.

THEORIES OF THE DRIFT.

319. No part of Physical Geology is so unsettled as that which relates to the dispersion of the Drift. It was formerly imputed to the agency of powerful currents of water alone, and by many it was ascribed to Noah's deluge. This was the view entertained by Dr. Buckland when he wrote his *Reliquiæ Diluvianæ*. But the short period of that flood and the absence of man's remains and works, with other considerations, have led to the universal abandonment of that view. The agency of water alone is now regarded as inadequate to account for the phenomena; they are attributed to the joint action of ice and water. Three theories are advocated by different writers on the subject.

320. The *iceberg theory*, supposes that the polar regions

of the earth were depressed beneath the ocean before the drift, and that during that period icebergs loaded with gravel and fragments of rocks were transported by currents, inflicting scratches and grooves upon the rocks over which they passed, and melting in lower latitudes deposited the materials they had conveyed.

It is urged in favor of this theory that such an agency is now witnessed in the phenomena of icebergs bearing thousands of tons of earth and rocks, transported by polar currents, as far as the drift extends. The stranding of the icebergs with fragments of rock frozen in them, is thought to produce striæ and furrows upon the bottom of the ocean like those of the drift.

It is objected to this theory that the boulders of icebergs are brought from high latitudes, while those of the drift appear to have been derived from neighboring mountains and ledges; the parallelism and uniformity of the striæ are not adequately accounted for; and the highest portions, as mountain tops, alone should have been furrowed and striated, since icebergs that would float over such heights would not reach to the bottom of the valleys, whereas the scratches are common in the plains and valleys.

321. The *elevation theory* attributes the phenomena to numerous elevations of the earth from beneath the polar seas, repeated for successive ages, sending enormous waves toward the equatorial regions bearing icebergs, with their boulders and earth, and urging before them the loose materials lying upon the surface; the striæ were produced by fragments of rocks in the bottoms of immense masses of ice which may have been forced up steep declivities.

The arguments in favor of this theory are, that

upheavals of the bottom of the ocean are known to occur, as in the elevation of Sabrina, Graham's Island, in the Mediterranean sea, and the Aleutian Islands, (§ 60;) that waves caused by earthquakes have great power to prostrate and transport heavy bodies; and that, as the surface of the continents, the mountain ridges and valleys, were essentially the same in form and direction then as now, the origin of the action must be sought for out of the country, and the direction of the drift dispersion indicates that the localities elevated were about the poles.

It is objected, that so numerous and extensive elevations as the theory requires, are improbable; that the phenomena of the drift indicate prolonged action, while such vertical movements are transient; and that the parallelism of the striæ on the rocks is not accounted for by moving fields of ice buoyed up by water, and conveyed by currents which, instead of passing up mountain sides and over their summits, would have swept around such obstacles.

322. The *glacier theory* supposes that the climate, which in the Tertiary period had been so warm as to allow the palms to grow within the temperate zones, became much colder, causing enormous sheets of ice—polar glaciers—to advance far beyond their previous limits, moving along the surface by alternate advance and retreat, rounding, polishing and striating the rocks, and afterward when melted depositing their loads of boulders and detritus, where the drift is now found. In Europe the center of expansion is supposed to have been the Scandinavian mountains, and in North America in the polar regions, from which the glaciers advanced southerly.

The advocates of this theory contend that the phenomena of glaciers as witnessed in the Alps (§ 32,) are perfect miniature representations of the drift—its striæ, furrows, boulders and moraines; that the elevation of extensive regions in high latitudes, like those of the Cordilleras in Mexico, and the high plains of Central Asia, would produce such a reduction of temperature as to cause immense glaciers, even thousands of feet in thickness. This theory is advocated by Prof. Agassiz.

The principal objection to the glacial theory is that glaciers are at present entirely confined to valleys, and the origin of such an enormous sheet of ice as it contemplates is altogether hypothetical.

323. Neither of these theories is deemed quite satisfactory; the proximate cause of the phenomena is very generally supposed to have been the joint action of ice and currents of water, but their origin and exact modes of operation are not determined.

THEORY OF VOLCANOES.

324. The cause of volcanic phenomena—eruptions, earthquakes, and elevations and depressions of portions of the earth's surface—has been the subject of much discussion. The prevalent theory on this subject supposes the whole earth, with the exception of a crust fifty or one hundred miles thick, to be in a melted state; that eruptions are produced by the access of water through crevices to this heated mass, which causes steam and other elastic bodies to force out through craters and fissures, lavas, scoriæ, sulphur, and other volcanic products; and that the whole globe has formerly been in a state of fusion, the present crust having resulted from the cooling of the surface.

325. In favor of this view it is urged—

1. That the *temperature* of the earth below a certain depth, as tested in mines and Artesian wells, continually *increases* as we descend (§ 5,) at an average rate of about, 1° for fifty feet, which would at a depth of a little more than a mile give the temperature of boiling water, and at a depth of about fifty miles, would be adequate to the melting of any known rock.

2. The *spheroidal form* of the earth is thought to indicate that it has been in a fluid state; and, if so, it must have been through the agency of heat. Sir Isaac Newton has shown that a body having the size and density of the earth, revolving on its axis with the rapidity it has, would, if its particles were free to move, assume its oblate spheroidal form.

3. The *numerous extensive volcanoes*, whose origin is deep seated, which communicate with each other over vast areas, and the masses of whose lavas thrown out at a single eruption sometimes surpass the bulk of the mountains in which their craters are situated, (§ 54,) require an enormous mass of heated matter; if the interior is in a melted state the materials are abundant, and their extrusion may be produced by the pressure of steam and other elastic bodies, or by the contraction of the crust upon the melted mass.

4. The phenomena of *hot springs*, *deep Artesian wells*, and the *increase* of *temperature* generally, as we descend beneath the surface, are adequately accounted for by this theory.

5. The *ultra tropical* character of the *climate*, and its great *uniformity* during the periods of deposition of the earliest fossiliferous rocks, have been attributed to this origin.

6. The rocks constituting the crust of the earth have been *melted;* the characters of the unstratified rocks show that they have undergone no change since they cooled from a state of fusion, and the stratified rocks consist of fragments of the unstratified, and have therefore been melted.

7. The phenomena of *earthquakes,* their great extent and violence, are accounted for under this theory, by the undulatory motion of the earth's crust in consequence of the expansion of gases within, or of undulations in the molten mass.

326. It is contended by objectors to this theory—

1. That the high temperature of the earth as we descend in it may be accounted for by chemical action, or by condensation of the air. To this it is replied that the phenomena occur where neither of these causes are adequate to their production.

2. It is objected that the temperature of the ocean is lower at great depths than at the surface. But strata of water arrange themselves in accordance with their specific gravities, the warmest rising to the top, and the crust of the earth is no thinner beneath the ocean than where it constitutes dry land.

3. Again it is objected that if the interior is intensely hot it should melt the crust with which it is in contact, or if not much hotter than the point of fusion at the time the crust consolidated, subsequent cooling should have caused it to solidify ere this. To which the advocates of the theory reply that the perfect non-conducting property of the crust prevents the escape of the heat. Baron Fourier has shown that the effect of this internal heat upon the surface is not the $\frac{1}{17}$ of a degree at present, and that the temperature has not fallen during the last two thousand years more than the

$\frac{1}{167}$ part of a degree. Currents of lava after accumulating a crust have been known to remain fluid within for many years, (§ 62.)

327. Another theory proposed to account for volcanic action, supposes there are extensive *repositories* of *melted rocks*, sufficient for the phenomena of volcanoes, while the great interior mass is solid; but most of the objections to the last theory are equally pertinent to this, and some of them apply with much greater force.

328. Sir Humphry Davy proposed to account for volcanic action by the hypothesis that the internal parts of the earth contain great masses of the metallic bases of the alkalies and earths—potassium, sodium, calcium, aluminium and magnesium—which on coming in contact with water decompose it, and produce vivid combustion. This hypothesis, though abandoned by Davy, has been advocated by others, and is not necessarily inconsistent with the doctrine of central heat; but the magnitude, universality and perpetuity of volcanic action indicate a more uniform and extensive source. By some Geologists, *electricity* is supposed to aid in the production of volcanic phenomena, and a remarkable concordance has been discerned between the prevailing direction of strata, and the curves of equal magnetic intensity, but our knowledge of its modes of operation is as yet quite imperfect.

329. The *gradual* elevation and depression of portions of the surface of the earth, as the rising of the coasts of the Baltic and the subsidence of Greenland, (§ 68,) are attributed to the expansion and contraction of the rocks in consequence of changes of temperature. It is ascertained by experiment that different rocks are expanded unequally by the same increase of temperature; granite less than

marble; marble less than slate; and slate less than sandstone. An increase of temperature of 600° applied to ten miles thickness of the earth's crust would elevate the surface two hundred feet; and a similar diminution of temperature would cause a corresponding subsidence.

330. The *elevation* of *mountain chains* is usually ascribed to violent volcanic uplifting agency; but some writers account for it by the collapse of the consolidated crust upon the contracting mass within, some portions of the crust rising in ridges, while others sink beneath its former level, thus increasing the relative height of the ridges. A modification of this view attributes the elevation to a *plication*, or folding of the strata, in consequence of horizontal or lateral pressure.

331. M. Elie de Beaumont contends that all mountain ranges which are parallel to each other, were elevated at the same time, even when situated remote from each other. The period at which a range was elevated is determined in accordance with principles illustrated in Fig. 62, § 103; for example, the Chalk series are found inclined upon the flanks of the Pyrenees, showing that they were deposited before those mountains were thrust up, while the lowest of the Tertiary rocks are not thrown out of their horizontal position; the epoch of the upheaval of the Pyrenees was therefore at the close of the Cretaceous, and antecedent to the Tertiary period. By the same test the Apennines are found to date from the same epoch, and the two ranges are nearly parallel. The same is found to be true of many other ranges at various periods, and hence Beaumont derives his generalization that parallelism indicates contemporaneity. He makes twelve systems of elevations in Europe, the first of which is the system of Westmoreland, in England, and

of the Hunsdruck on the Continent, which was thrown up during the Silurian period, being the oldest upheaval as yet identified upon the globe; and the last is the system of the Principal Chain of the Alps, which was elevated after the close of the Tertiary period, being the last great convulsion to which Europe has been subjected. Five or six systems have been assigned to the American Continent, of which that of the Andes is the most recent, its upheaval having occurred, as Beaumont supposes, in the historical period. These views, though generally received with favor by Geologists, require confirmation; if they can be shown to be correct, their guidance will greatly facilitate geological research.

THEORY OF VEINS.

332. It is generally supposed that most veins were *injected* in a fluid state into fissures in both stratified and unstratified rocks; some veins have been traced to large masses of the same materials, whose former fluidity through the agency of heat is deemed demonstrable.

It is however admitted that some veins are contemporaneous with the rocks in which they are included, having been separated by chemical *segregation*, (§ 106,) as the flints were separated from the chalk, and the garnets from the mica-slate; in such cases they are entirely included in the rock.

The materials of some veins appear to have been *sublimed* into fissures, studding the interior with crystals. An experiment has been instituted to test the correctness of this view of the origin of veins. Lead ore (Galena) was sublimed through steam of water in an earthen tube, and condensed in cubical crystals in the colder parts of the tube; boracic acid was sublimed and condensed in the same manner.

333. By some Geologists veins are supposed to have been formed by the chemical changes that may have taken place under the influence of *electrical* currents in the interior of the earth; the experiments of M. Becquerel on the insoluble compounds of copper, lead, and lime, show that many crystallized bodies, hitherto found only in nature, may be artificially formed by the long-continued action of very feeble electrical currents, and Mr. Fox accounts for the superior richness of metallic veins running east and west, by the electro-magnetic currents circulating in that direction, decomposing metallic compounds and transferring their elements to a considerable distance in the rocks.

COSMOGONY.

334. The Science of Geology does not furnish the means of determining in what state the materials of the earth were when created, nor can it assure us that those materials have not undergone many important changes, of which we have no indications. Conjectures, however, have been formed respecting the earliest conditions of created matter. A prevalent *hypothesis* supposes matter to have been created in its elementary forms; chemical attraction caused many of these elements to combine with each other; this rapid chemical action—combustion—evolved sufficient heat to vaporize a large portion of the substances; subsequent radiation condensed the vapors to a liquid state, and a solid crust accumulated upon which the waters of the atmosphere were precipitated, bringing the phenomena within the province of positive Geology.

335. The plausibility of this hypothesis is argued from

its accordance with the known laws of Chemistry, and Astronomical analogies. Comets and some nebulæ present matter in an exceedingly attenuated state, so that stars can be seen through the former in some instances with scarcely any diminution of lustre; and some of them, it is thought, are becoming gradually more dense at their centers. The moon presents the appearance of a globe with its surface shaped by igneous agency. Its mountain peaks rise more than four miles in height, and some of its volcanic craters, which are one hundred and fifty miles in diameter, have their bottoms depressed more than twenty thousand feet below the general surface. These craters very closely resemble in form, terraces, etc., some of the earth's volcanoes, especially Kilauea, (§ 58.) As there is no water upon the moon, and a very rare atmosphere, if any, no stratified deposits exist, nor are the characteristic effects of volcanic action obliterated, but it presents to us such an appearance as Geologists assign to the earth in the primary period, before the agency of water modified its surface by wearing the igneous rocks and depositing strata of the detritus.

CHAPTER XI.

PRACTICAL GEOLOGY.

336. By *Practical* or *Economical Geology* is understood an exhibition of the facts of the science obtained by observation, and the laws deduced from the facts by generalization, with reference to their immediate application to the wants of society. Its importance can not indeed be adequately estimated by monetary tables, since its effects on mind in stimulating intellectual activity, and inducing wholesome mental discipline, are no less valuable than what it has accomplished for the comforts of society and the interests of commerce. But its cash value is distinctly appreciable. Mr. Miller states, in his "Old Red Sandstone," that the time and money squandered in Great Britain alone in searching for coal in districts where the well-informed Geologist could have at once pronounced the search hopeless, would much more than cover the expense at which geological research has been prosecuted throughout the world. The Old Red Sandstone, the Silurian rocks, and talcose and mica-slates have been bored for coal, where the author just quoted remarks, "there might be some possibility of penetrating to the central fire, but none whatever of reaching a vein of coal."

337. Not only is the physical condition of a country influenced by its geological structure, but the occupations

and habits of its inhabitants are almost exclusively determined by it. A good geological map of a country is the best index of the relative values of its districts for particular economical purposes.

Practical Geology relates particularly to the processes of *Mining*, *Engineering* and *Architecture*, and *Agriculture;* it depends principally upon the fact that minerals which are useful for practical purposes are found only in certain geological formations.

338. In *mining* for most *metals* Geology indicates the *primary* and *metamorphic* rocks, the junction of the stratified with the unstratified rocks, as the most promising fields of search.

Gold and *Platinum* are found not as *ores*, but *native metals* in quartz rock and talcose slate. They are however obtained from the drift and alluvium, which consist of the detritus of these primary rocks.

Silver occurs as a sulphuret and a chloride in the primary and transition slates; it is also associated with metallic copper.

The principal ore of *Mercury*—the sulphuret—is found in mica-slate and in the New Red sandstone.

The ores of *Copper*—the sulphuret, the oxide, and carbonate—are found in the primary rocks, and in connection with trap dikes in the secondary; copper is also found native in these situations.

Lead ore is the sulphuret—galena; the unstratified and stratified primary rocks, the metaliferous limestone, and the secondary rocks as high as the lias, are its repositories.

The sulphuret of *Zinc*—zinc-blende—and the carbonate of zinc, are found in the Transition and Secondary series.

Tin, Antimony, Bismuth, Cobalt, Arsenic, Manganese, etc., are found in the oldest rocks.

Iron occurs in all formations, in quantities and forms adapted to working, but the iron ores of the older rocks are the most valuable.

The metals are usually found in veins, but sometimes they constitute true beds, and are also diffused in fragments through a rock, as in the drift, or alluvium. Veins follow certain courses relatively to the principal axes of elevation of the country; they are often interrupted by cross veins and dikes, and thrown either up or down. The veins containing metallic ores are called *lodes;* those not metaliferous, *cross courses.* The inclination of the vein to the horizon is called its *underlie, hade* or *slope;* and its intersection with the surface, its *direction.* The practical Geologist is enabled to determine these, and map them out so as to guide the miner to the readiest and most economical method of developing the ore.

339. True *bituminous coal* is found only in the carboniferous system of rocks, and in a certain part of the system only; all search for it in other positions has proved fruitless. Beds of lignite do occur in the Oolite, Lias, and Tertiary, but they are rarely worth working. *Anthracite,* which is coal deprived of bitumen, is sometimes found below the carboniferous series. Masses of bituminous matter sometimes occur in the Old Red sandstone with markings so similar to the vegetable impressions on the carboniferous sandstones, as to deceive an unpractised eye; but the groups of fossils are characteristic of the formations and enable the Geologist to discriminate them. Dislocations of beds of coal and strata associated with them are singularly frequent, and the *faults* are so complicate as to require

much geological knowledge and experience to aid the miner in recovering his lost seams. The most successful modes of draining and ventilating the mines are also indicated by the practical Geologist.

340. The *diamond*, which is pure crystallized carbon, and is supposed by many Geologists to be of vegetable origin, has been found in the talcose slate, and the New Red sandstone; the rocks with which it was associated indicating the agency of heat. The precious stones, *emerald, ruby, sapphire, topaz, carnelian, tourmaline, garnet*, etc., are found in the igneous unstratified rocks. The gems are frequently obtained from the drift, together with gold and platinum, having been removed from their original positions by the abrasions of the older rocks.

341. A knowledge of Geology is highly important to the engineer and architect. "It is in proportion to his acquaintance with Geology and Mineralogy," says Mr. Cresy, "that the Civil Engineer is rendered skilful in the formation of roads, canals, harbors, building of bridges, or forming foundations of any kind, and draining; wherever the scene of his labors may lie, he can not be entirely successful, without a careful consideration of the various strata composing the earth's crust. When Smeaton was called upon to construct the Eddystone Lighthouse, he commenced by examining the structure of the rock on which it was to be based, and as far as possible to endeavor to imitate nature in his arrangement of the courses. Had the builders of the Leaning Tower, at Pisa, been equally careful, or had they been acquainted with the composition of the earth on which they laid their foundations, the world would never have had the opportunity of supposing that its inclination was the effect of design, instead of the consequence of an

insecure base, which might have been consolidated by art; had the alluvial matter on which the footings are laid been converted into a mass of conglomerate or artificial rock, this famed Campanile would have stood as upright as the Eddystone Lighthouse. For all the purposes of building, it is necessary that the constructor should be acquainted with the formation and properties of the matter with which he has to deal; he should understand the cause of the durability of a substance, whatever it may be, as well as what disintegrates or destroys it."*

A lithological map of a country delineating its various strata, their dip and strike, is an efficient guide to the Engineer in locating public works, estimating their expense, and procuring the materials for their construction.

342. One of the most important uses to which minerals are applied, is for architectural purposes. The character of the material requires to be adapted to the circumstances in which it is used, in order to secure durability; some rocks disintegrate rapidly in consequence of expansion and contraction from changes of temperature, and others are destroyed by absorbing water from the atmosphere, which expands in freezing. The importance of the durability of materials has been much neglected in modern architecture. Mr. Ure remarks that, "such was the care of the ancients to provide strong and durable materials for their public edifices that but for the desolating hands of modern barbarians, in peace and in war, most of the temples and other public monuments of Greece and Rome would have remained perfect at the present day, uninjured by the elements during two thousand years. The contrast in this respect of the works of modern architects, especially

* Encyclopædia of Civil Engineering, by Edward Cresy, p. 617

in Great Britain, is very humiliating to those who boast so loudly of social advancement; for there is scarcely a public building of recent date which will be in existence a thousand years hence." This frailty of public structures is equally conspicuous in the United States.

343. *Granite, Syenite* and *Porphyry* are valuable materials for building. To adapt them for this use they should be fine and uniform in texture, as the coarser varieties are not so coherent. They should be free from metallic bodies, especially iron pyrites—sulphuret of iron—which, on exposure to moist air, rust, discolor, and disintegrate the rock in which they are embedded. These rocks usually harden after removal from the quarry and exposure to the air; their geological position is in the Primary series, but they occur intruded as dikes and veins in the more recent strata.

The varieties of *Trap* and *Basalt* are used for building; the natural faces of the basaltic pillars require no dressing to fit them for the purpose, and the sombre hue of the ferruginous varieties is adapted to some styles of architecture. These igneous rocks are associated with the Secondary series of strata.

The *Lavas* of the Tertiary and the Alluvial periods, are also used, but are frequently not sufficiently firm and compact for this purpose.

344. *Sandstones* usually consist of grains of quartz, with some admixture of other minerals, as feldspar and mica; some are cemented together with carbonate of lime, and are called *calcareous*, others with clay, and are denominated *argillaceous*. Their colors, yellow, brown, red or black, are principally due to the compounds of iron. They occur in strata of all the geological series, and are, there-

fore, widely diffused and easily accessible. They are extensively employed, frequently under the name of freestones. To fit them for this purpose they should be free from iron-pyrites, iron sand, or any substance which will on exposure to the weather undergo chemical change. Some sandstones, are worthless for ordinary building purposes, falling to pieces as soon as they dry, but are very firm while kept beneath the surface of water or in the ground. Sandstones which absorb much water will not bear exposure to frost. This may be tested by immersing them in water and exposing to frost, or in a saturated solution of sulphate of soda—Glauber's Salts—and drying in the air; in the latter case, if much of the solution is absorbed the crystallization of the salt will produce the same disintegrating effect as the frost would.

The *Conglomerates* and *Breccias* are not so well adapted to architectural purposes as the rocks which are finer and more uniform in structure.

345. *Limestones* have always been highly esteemed for building and various other purposes. They are very abundant, occurring in all the series from the statuary marble of the primary to the marls of the alluvial. They are either *granular* or *compact* in texture; the former furnishing the firmest and finest marbles. When pure they are white, but they are often clouded with black mica or some metallic compound. These are found in the vicinity of the igneous rocks, to which they are supposed to owe their crystalline texture. The finer varieties of marble are said not to be exceeded in durability by any other rock used in architecture. Some of the compact varieties of limestone are easily wrought, are susceptible of polish, and are well adapted to building purposes.

Slates are much used for roofing. They should be homogeneous and fine in texture, of uniform cleavage, free from pyrites, impermeable to water, and sufficiently tenacious to allow perforations for nailing them. Such slates are found in the older stratified series of rocks.

346. Minerals useful for other purposes are found in various rocks. *Sulphate of Lime* or *Gypsum*, extensively used for forming stucco, taking casts, etc., is found in the Transition, Secondary, and Tertiary series.

Steatite or soap-stone, employed as fire-stones to line furnaces and stoves, is soft, may be sawn, and turned in a lathe; a compact variety of it, called pot-stone, is wrought into culinary vessels in Italy. This rock occurs among the oldest strata.

Salt—chloride of sodium—is frequently found with gypsum in the New Red sandstone, but occurs also in other strata. The remarkable deposits in Poland and Hungary are in the Cretaceous or Upper Secondary; the mines in Poland have been worked since A. D. 1251, and are estimated to contain sufficient salt to supply the world for many centuries. Hills of salt three hundred to four hundred feet high are found in the Cretaceous strata at Cardona in Spain; but the Catalonian deposits are in the Tertiary. Some salt lakes are at the present day producing deposits by evaporation; as exemplified by the Dead and Caspian seas, the lakes of Northern Africa, and the Great Salt Lake, which is situated upon the flanks of the Rocky Mountains at an elevation of four thousand two hundred feet above the sea, with an area of two thousand square miles. The brines from which most of the salt in the United States is obtained, come from below the coal. Forty gallons of the New York springs yield a bushel of salt. Rock salt has been found in Virginia and in Oregon.

347. *Clay*, consisting of alumina and silica, owes its plasticity to the former ingredient; it results from the disintegration of feldspar and slate rocks, and is found principally in the Tertiary and Drift. It is often mixed with the carbonate of lime, magnesia, and the oxide of iron. Clay used for making bricks, generally contains a portion of the hydrated oxide of iron, which is decomposed when heated, forming red oxide of iron, and imparting its color to the brick; some clays, however, contain no iron, and the bricks made of them are of a light color, as is the case with those made at Milwaukie, Wisconsin. Clay for fire-bricks should contain no iron, magnesia or lime, as those ingredients impart fusibility; such clay is found in the Tertiary and in the Coal formation. *Pipe-clay* or *Potter's clay* are pure varieties, and consequently white. *Porcelain clay* or *kaolin* is decomposing feldspar, which mixed with silica, lime and unchanged feldspar, produces beautiful specimens of earthenware, some of which are translucent porcelain. The kaolin occurs in extensive beds in granite rocks. The Sevres ware made in France consists of sixty-five parts of kaolin, twenty of feldspar, ten of flint or quartz, and five of chalk. The China ware contains more quartz and is more glassy.

Fuller's earth is composed of silica, alumina, lime, magnesia and the oxide of iron; it has a soapy feel and is used for removing grease from woolen cloths.

348. *Sand* is usually grains of quartz, mixed with grains of mica, feldspar, oxide of iron, etc. Sand is extensively used in the manufacture of glass, which is a transparent, fusible compound of quartz—silicic acid—and

potash or soda; the oxide of lead and lime are sometimes added for glass of different kinds. Sand for the manufacture of glass should be pure, especially free from the oxides of metals, which impart to it deep colors; the color resulting from one metal may sometimes be discharged by another, as the green color resulting from the oxide of iron is removed by the oxide of manganese. Beds of sand adapted to this purpose are found in the Tertiary, Drift and Alluvium. That which is obtained by pulverizing sandstones is apt to contain a troublesome amount of iron, or other impurities. Sand used for mortar should be fine and of sharp grit.

349. *Quicklime*, used for mortar, for purifying coal-gas and syrups, for fertilizing land, and for various other purposes, is obtained from limestones, by expelling the carbonic acid by heat. The purest marbles furnish the best lime, but some impurities are not detrimental to it for certain purposes. The strength of mortar depends upon the formation of the chemical compound of lime, silica, and water, and is not necessarily impaired by the presence of a small portion of iron, clay, &c. The rich or *fat* limes double their volume in slaking, and absorb nearly three hundred per cent. of their weight of water; the *poor* limes augment their volume slightly in slaking, and absorb about two hundred per cent. of their weight of water.

Hydraulic lime is distinguished by forming a mortar which sets under water, and consists of silica, lime, alumina, magnesia, and frequently the oxide of iron; the last ingredient is deemed undesirable for most purposes to which this lime is applied. Prof. Beck gives the composition of a variety of this substance extensively used in the State of New York as follows: carbonic acid 34 20, lime 25.50, magnesia 12.35, silica 15.37, alumina 9.13, and peroxide of iron 2.25.* Hydraulic limes are not uniform in their composition.

Parker's Cement, formerly patented in England, consisted of fifty-five parts of lime, thirty-eight of alumina, and seven of oxide of iron. It was obtained from the *septaria* found in the argillaceous strata of the Oolite and the Tertiary. Other septaria forming excellent cements differ in the proportions of their ingredients. Greater caution is requisite in burning hydraulic lime, since it is fusible, and the heat applied to the common lime will vitrify this substance and render the process quite imperfect: common lime will bear a white heat, but the calcination of hydraulic lime is not well effected above a red heat.

Puzzuolana, a volcanic tufa composed of silica, alumina, lime, magnesia, soda and oxides of several metals, forms with lime and water a strong hydraulic cement. The strong cement used in the construction of the Eddystone lighthouse was made of equal measures of puzzuolana and blue lias lime, slaked into a powder; it set slowly but very firmly under water. The ancient Romans used puzzuolana in their mortars, but the cement of their structures found in England and other parts of Europe, whose hardness after the lapse of centuries excites the admiration of architects, consists of lime, sand, pounded brick or tile-dust, and wood ashes; it is of a reddish color, and contains cavities lined with crystals of carbonate of lime. Artificial puzzuolanas are made by mixing clay with lime; pipe clay and lime after burning will set into a very firm cement.

350. *Buhrstone*, used almost exclusively for millstones, is a cellular variety of quartz, and owes its value for this

* Report on the Mineralogy of New York, p. 78.

purpose to the hardness and sharpness of the inequalities of its surfaces. The finest stones have usually been imported from France where they were found in the Tertiary of the Paris basin, but stones of excellent quality are obtained in Muskingum county and other localities in Ohio, where it is associated with the carboniferous sandstones. As some of the cavities contain lime, it is conjectured that the removal of that substance by solution has produced the cells. It is found also in Georgia and in Arkansas.

351. *Marls* are composed of clay and lime, are very variable in their constitution, and their value for fertilizing soils depends upon the circumstances under which they occur. Their calcareous contents have oftentimes been derived from the shells or other organic bodies embedded in them. They are sometimes colored blue by the protoxide of iron, and red or yellow by the peroxide of the same metal. They are found in all parts of the series of stratified rocks; as the older marls are highly indurated, those which are available are confined to the Tertiary and Alluvium. The principal deposits are beneath ponds and peat swamps. The name is sometimes improperly applied to the greensand of the Cretaceous formation, which owes its fertilizing quality to the alkalies it contains, and not to lime or clay; but the chalk marl is very largely calcareous.

352. Geology is second only to Chemistry in the amount and importance of the aid it renders to scientific *Agriculture*. "The geologist," says Prof. Johnston in his Lectures on Agriculture, "can best explain the immediate origin of the several soils The cause of the diversities which even in the same farm, it may be in the same field, they not unfrequently exhibit; the nature and differences among

the subsoils, and the advantages which may be expected from breaking them up or bringing them to the surface." Intelligent practice of the art of Husbandry is based upon a knowledge of the constituents—organic and inorganic—of the crops to be raised, and of the ingredients of the soil, which are also both organic and inorganic. These must be adapted to each other. All plants derive the materials of their growth from the earth and atmosphere; they have not the power of creating any element, but their chemical composition is a perfect index of their food. All animals obtain their subsistence from vegetables, or from other animals which subsisted upon the products of vegetation, so that all organic matter is derived from the inorganic—mineral—kingdom, through the medium of vegetation. After death all animal and vegetable substances by putrefaction return to the earth and atmosphere, from which they originated, to be again absorbed by growing plants and furnish food for animals; thus exemplifying that great principle of instability by which the stability of nature is secured—that cycle, the mineral, vegetable and animal, through which particles of matter are incessantly passing.

353. The inorganic part of soils consists of two classes of substances—the *salts*, which are soluble in water, from which plants obtain their saline matters, and which constitute their ashes when burnt; and the insoluble *earths*, which form the great bulk of most soils.

The principal soluble saline ingredients of soils, are iron, sulphates of lime magnesia and soda, nitrates of potassa soda and lime, phosphates of lime and magnesia, and chloride of sodium—common salt. The carbonates of lime magnesia and iron, are soluble in water charged with carbonic acid. The various saline substances are not all found in the same soil: one soil may contain soda and be deficient in the salts of lime; another may be supplied with the phosphates and be destitute of the sulphates. If some of these ingredients abound and the others are deficient, the soil will be favorable to the growth of certain plants, and unfavorable to others. The amount of a salt in a soil is oftentimes very small, and might be deemed too insignificant to furnish food to plants, but a single grain of saline matter in every pound of soil a foot deep, is equal to five hundred pounds to the acre. The salt, found in a plant, however minute the quantity, is indispensable to its growth—without it the soil is for that plant sterile. These salts are derived from the mineral masses of the soil. The water percolating through such soils holds them in solution; they decompose soaps, and hence are said to render the water *hard*.

354. The other class of inorganic constituents of soils is the *earthy*, making ordinarily ninety per cent. or more, of the whole. It embraces three principal ingredients. *Silica*, in the form of sand; *alumina*, mixed or combined with sand as *clay*; and *lime* principally in the form of *carbonate*, as limestone or chalk. Soils are named according to the proportions in which these are mingled. According to Johnston, one hundred grains of dry ordinary soil, containing only ten of clay, form a *sandy* soil; ten to forty grains of clay make a *sandy loam;* forty to seventy, a *loamy soil;* seventy to eighty-five a *clay loam;* and from eighty-five to ninety-five a *strong clay*, fit for bricks or tiles. Arable land rarely contains more than from thirty to thirty-five per cent. of alumina. If a soil contain ten per cent. of the carbonate of lime, it is called *calcareous*, and if it has more than twenty per cent. of it, it is designated a *marl*.

The oxide of iron forms two or three per cent. of sandy soils, and in red soils much more.

Silica presents itself in soils as sand, small fragments of the mineral quartz, one of the constituents of the granite rocks; or combined with alkaline and earthy bases, as silicates of potassa, lime, alumina, etc. The feldspar of granite yields it in the form of silicates of alumina and potassa. The silicates are slightly soluble, dissolving no faster than the necessities of plants demand, hence they are not rapidly washed out of the soils, as are some soluble ingredients. Silica is found in soils in variable proportions, but usually predominates over all the other constituents. Silica is contained in the stems of plants, especially of the grasses.

Clay consists of the silicate of *alumina*, mixed with uncombined alumina and silica. It is derived from the abrasion of the slate rocks, and from the feldspar of granite. Its most striking property is its adhesiveness; soils are close and compact in proportion to the quantity of clay they contain. When it predominates, it constitutes the heavy cold clay land, requiring under-draining, and sometimes the addition of sand to render it fertile. It retains manures well, being almost impervious to water. Clay soils exhale a peculiar odor called argillaceous, when they are breathed upon.

Lime exists in soils in very variable quantities, as fragments of limestone disseminated with the other materials, or held suspended or dissolved in water, as the carbonate, sulphate, or phosphate of lime. Lime is not found pure or caustic in soils, and if it is applied in this state, it soon loses its causticity by the neutralizing power of the various acids it encounters.

355. The *subsoil* is of various characters: in some cases consisting of porous sand or gravel, in others of a light loam or a stiff clay. A stratum of these materials, of variable thickness cemented by the salts of lime and iron, and indurated, is called the *hardpan*.

356. The *organic* portion of soils consists of the animal and vegetable substances found in them, which are either the relics of ancient animated forms buried in the rocks, or more frequently the decaying bodies of plants and animals of the present period. Geology leads us to the presumption that in the early history of our globe, mineral matter existed alone, and that subsequently the Creator introduced various races of plants, which drew the elements of their matter from the mineral kingdom. The plants first introduced were simple in their organization, and capable of living upon the elements of air, water and mineral salts, without any previously organized matter for their nutrition. They were cryptogamous, flowerless plants, like the sea weeds and mosses of the present day. These plants having grown and died, furnished by their decay the organic matter necessary for the nutrition of the more highly organized flowering plants, subsequently introduced. The plants which are cultivated in Agriculture are the flowering ones, for these only bear seeds; soils therefore must contain some portion of organic matter for their nutriment. The quantity however varies greatly: in peat soils, it forms from fifty to seventy per cent. of the whole weight; in rich meadows it may amount to twenty per cent., but the proportion is generally much smaller.

Oats and rye will grow in a soil which contains only one or two per cent. of organic matter; barley with three per cent.; while good wheat soils require from four to eight per cent. Organic matter alone will not produce a fertile soil; the inorganic, earthy and saline ingredients must be present. It must also have undergone decomposition: if it is secluded from the oxygen of the air it will not decay and yield up its elements the appropriate nutriment to surrounding plants; a peat bog covered with stagnant water yields very little nutriment to other plants, but exposed to the air and mixed with ashes or other substances yielding alkalies, it becomes a very efficient fertilizer.

357. The *origin of all soils* is in the disintegration and decomposition of rocks, produced by the mechanical and chemical agencies of water, air, etc. The amount of soils furnished by groups of rocks depends upon the composition and structure of the rocks. Many slates and shaly limestones and sandstones disintegrate rapidly, through the influence of water and frost penetrating between their laminæ. Limestones suffer constant loss of materials by the solvent power of rain water holding carbonic acid in solution. The presence of alkalies in the feldspar and mica of granite and gneiss, greatly facilitate the disintegration and decomposition of those rocks. Calciferous sandstones are liable to decay by the solution of their lime leaving the sand; much of the lime passes through the soil and accumulates in the subsoil or hardpan. Rocks which contain metals or metallic compounds which readily suffer chemical change by exposure to the air, are disintegrated with great rapidity: this may be exemplified by *pyrites*—sulphuret of iron—which is frequently found in rocks especially the slates; the sulphur unites with the oxygen of the air, forming sulphuric acid, and the iron with oxygen producing the oxide of iron, which by combination with the acid forms sulphate of iron or copperas. The sulphate

of iron dissolving in water, the rock becomes porous and crumbles; flowing over limestone rocks it decomposes them, producing sulphate of lime, or gypsum; or if it comes in contact with the feldspar of granite, it decomposes that mineral, forming with its potassa the sulphate of potassa, depositing its oxide of iron in the form of iron rust.

The *friction* of falling water wears the rocks which are subjected to its agency.

Vegetables undergoing decomposition generate acids, which act chemically upon the rocks, forming soluble salts with their alkalies and earths: living vegetables also exert a powerful influence upon the rocks; mosses and lichens growing upon bare granite rocks, absorbing their soluble parts, decompose and disintegrate them.

358. Over extensive areas the soil is derived directly from the rock upon which it rests; it however differs somewhat from the rock in composition; analysis shows that fragments of rocks exposed to atmospheric influence lose a part of their soluble matter, so that their debris is composed of a larger proportion of insoluble parts. But great masses of soil have been transported from the localities in which they were formed by the agency of currents of water, either *alluvial*, or exerted on a large scale in the *drift*, commingling the soils from various rocks, so as to render the earth more uniformly fertile by a mixture of various ingredients, which is known to produce a favorable result. Drift soils are recognized by their heterogeneous character, and by their pebbles, sometimes called cobblestones, consisting of water-worn fragments of the hardest rocks.

359. Soils are usually classified in accordance with the

predominance of some element, as sandy, argillaceous, loam, etc.; but some writers base their classification on their fitness for certain crops, as wheat districts, corn districts, etc., or on the geological formations, in which case each district is underlaid by characteristic rocks. Prof. Emmons, in his Report on the Agriculture of the State of New York, divides the State into six districts coinciding with six groups of rocks, which impart to the soils, in a good measure, their distinguishing characters.

360. Soils derived from *granite* and *gneiss* contain the earths and salts requisite for a high degree of fertility, but they are often too silicious and porous, and their value depends upon their position and the nature of the subsoil.

Syenitic and *Hornblende* soils contain, in addition to the usual constituents a large proportion of the oxide of iron, magnesia, and the oxide of manganese, and are quite fertile.

The soils derived from *Trap*, *Greenstone* and *Basaltic* rocks also contain a large per centage of lime, magnesia, and iron, and are highly fertile. The *Lava* soils, whether trachytic or augitic, owe their remarkable fertility to the large quantity of alkaline salts they contain.

The *slates* produce a variety of soils, in many places thin and poor, but in others deep and capable of being made good.

Calcareous or *limestone* soils are also very variable in their quality; those which contain magnesia or iron are fertile.

Sandstone soils require the admixture of other substances, especially clay, to make them adhesive and fertile.

Alluvial soils are generally rich, consisting of fine divided matter thoroughly commingled; they are found the mouths of rivers, and in their valley , and are frequently called *bottom lands*.

361. The art of maintaining an uninterrupted fertility, or renovating an exhausted soil, consists in supplying those ingredients, mineral and organic, which have been removed in the crops. The earthy and saline substances may frequently be obtained by subsoil ploughing, thus forming the earths and salts by the agency of the oxygen of the air upon the insoluble minerals of the subsoil; or they may be imported from other localities, and applied as mineral manures or fertilizers, such as gypsum, marl, or lime. Many decomposable shales containing iron pyrites, mixed with lime produce gypsum, and constitute excellent fertilizers. A deficiency of organic matter must be supplied by turning into the soil green crops which have drawn much of their carbon from the carbonic acid of the air, or by the application of decaying animal or vegetable substances obtained from other sources.

362. *Drainage* is indispensable to successful agriculture where the water of stiff clay lands or swamps is retained by impervious subsoils, inclined strata, or dikes. Geology aids in effecting the drainage either superficial or deep, by indicating the general laws which appertain to dikes and strata, and their permeability to water.

363. Not only is the physiognomy of a country influenced by the different outlines which the various geological formations impart to it, but its *scenery* is greatly modified by the different kinds of vegetation, whether indigenous or induced by cultivation, which the various soils sustain.

CHAPTER XII.

THE HISTORY OF GEOLOGY.

364. GEOLOGY is one of the most recent branches of physical science: but little more than half a century has elapsed since it was elevated from the state of absurd hypothesis and wild speculation to the rank of an inductive science, based upon accurate observation of facts. The earlier cosmogonies were altogether fanciful, and in some instances ridiculous. Some of the ancient philosophers, however, appear to have apprehended correctly the origin of many geological phenomena. Pythagoras recognised the operation of the existing causes of change on the earth, as the wearing away of the coasts, and the formation of alluvial deposits, by marine currents and waves: the geographer, Strabo, was convinced, by finding fossil shells far above the sea level, that parts of the earth had been raised by volcanic agency.

365. The numerous perfect fossils of Italy early excited a spirit of inquiry respecting their origin. From the commencement of the sixteenth century, two questions regarding them were discussed at great length, viz., first, Whether they ever belonged to living beings, or were mere semblances of animals and plants; and secondly, If they did, whether they were overwhelmed and imbedded in the rocks by the deluge of Noah. By some writers, fossils were thought to be *"lusus naturæ,"* (sports of nature,) the results

of the operations of a *materia pinguis*, "fatty matter" found in some parts of the earth, fermented by heat; and others, in an equally unintelligible mode, ascribed them to "tumultuous movements of terrestrial exhalations." These unprofitable discussions were continued for more than two centuries, and in England assumed a theological cast, some writers contending that the Scriptures contain a perfect system of natural philosophy in detail. One of the treatises characteristic of the age was Burnet's "Sacred Theory of the Earth; containing an account of the original of the Earth, and of all the general changes which it hath already undergone, or is to undergo till the consummation of all things," published in A. D. 1690. Of this work Sir C. Lyell remarks, "Even Milton had scarcely ventured in his poem to indulge his imagination so freely in painting scenes of the Creation and Deluge, Paradise and Chaos." It was, however, at the time regarded as a work of profound science.*

366. In A. D. 1775, Werner, a professor of mineralogy in the School of Mines, in Germany, commenced teaching that all rocks, unstratified as well as stratified, were deposited by water, that all formations were universal, and that veins were filled by precipitation from aqueous solutions. This was denominated the Neptunian theory, from Neptune, the god of the sea. About the same time, Hutton, a geologist of Edinburgh, published his theory of the earth, ascribing the origin of all rocks to fire or heat. The melted rocks, after consolidation, he supposed, were abraded by the action

* For an admirable sketch of ancient cosmogonies and "Theories of the Earth," consult "Lyell's Principles of Geology," Book I, Chapters 1—4.

of water, and deposited as strata, through which igneous rocks often protruded themselves, producing the effects of heat on the surrounding masses. Continents, he supposed, were elevated by volcanic agency, and veins filled by injection of melted matter from beneath. These alternations of fusion, consolidation, abrasion, and deposition, may be repeated, and geology gives no intimation of the time when the series of changes commenced. This view of Hutton was designated the Plutonian theory, from Pluto, the god of fire. While geologists were discussing the merits of these rival hypotheses, Mr. William Smith, an English surveyor, having explored the whole country on foot, without the guidance of previous observers, published his "Tabular View and Map of England," in which he exhibited the order of the various strata and the relations of their fossils, thus accomplishing more for science than had been effected by centuries of discussions.

367. The excess of theorizing on the subject induced distrust of all systems, and led the Geological Society of London, formed in 1807, to devote itself to the accumulation of accurate observations: the success attending these efforts soon rescued the science from the imputation of being a visionary pursuit. The recent rapid advancement of geology is due in no small degree to the progress of the collateral branches of science—botany, zoology, and comparative anatomy—since the palæontological characters of rocks are much more reliable indications of identity or diversity than the mineral characters.

368. In America, Mr. Maclure, having explored a large part of the United States, in 1810, published his "Observations on the Geology of the United States," giving the first sketch of the distribution of the stratified

rocks of this country, referred to the European standard. Professor Eaton surveyed the rocks on the line of the Erie canal, through the State of New York, and published the results in 1824. The same year commenced a series of geological surveys of states by legislative sanction, which has progressed until two-thirds of all the states have been completely or partially surveyed. Besides accomplishing the primary design in developing the natural resources of the states, these surveys have accumulated an immense array of accurate observations for the general advancement of science. By comparison and generalization of these results, a highly satisfactory view of the geology of this country may be obtained. The spirit of independence of European classifications, and the determination to develop the rocks as they are, rather than to identify them with the sub-divisions of foreign systems, which have characterised these surveys, have obviated some of the hindrances which retarded the progress of the science in this country.

369. Although Geology is by no means complete, since in no science are facts more rapidly accumulating, and theories and systems are but expositions of the present amount of knowledge on the subject, still its immense array of facts and legitimate deductions constitute a science as well established as is Chemistry or Astronomy.

CHAPTER XIII.

RELATION OF GEOLOGY TO RELIGION.

370. It is not customary in elementary treatises on branches of physical science to exhibit their relations to morals or religion, since it is the office of other branches of human learning, as natural theology, to treat specifically of those relations in detail. All science has such relations, since it is an exhibition of the laws which the Creator has established over matter, illustrative of his power, wisdom, and benevolence, and religion consists in a knowledge of the Creator and the exercise of those affections which such knowledge enjoins. The science of Geology has, through misapprehension of facts and opinions, been regarded with jealousy, as favorable to infidelity and even atheism, teaching the eternity of matter, its self-guiding and renovating power, and dispensing with a Deity in the creation and regulation of the world.

371. On the contrary, those who, with competent knowledge of the science of geology and the art of interpretation, have carefully examined this subject, confidently assert that no other science furnishes an equal number of striking illustrations of natural and revealed religion.

"Shall it then any longer be said," says Dr. Buckland, "that a science, which unfolds such abundant evidence of the being and attributes of God, can reasonably be viewed in any other light than as the efficient auxiliary and handmaid of religion? Some few there still may be, whom timidity or prejudice, or want of opportunity, allow not to examine its evidence; who are alarmed by the novelty, or surprised by the extent and magnitude of the views which geology forces on their attention, and who would rather have kept closed the volume of witness, which has been sealed up for ages, beneath the surface of the earth, than impose upon the student in natural theology the duty of studying its contents; a duty in which, for lack of experience, they may anticipate a hazardous or a laborious task, but which, by those engaged in it, is found to afford a rational, righteous, and delightful exercise of their highest faculties, in multiplying the evidences of the existence, attributes and providence of God."

"Let not the Christian divine refuse the aid offered by physical science. Let him no longer indulge groundless jealousies against true philosophy, as if adverse to religion Especially let him not spurn the aid of geology, which alone of all the sciences, discloses stupendous miracles of creation, in early times, and thus removes all presumption against the miracles of Christianity and special providence at any time. It is indeed an instructive fact that a science which has been thought so full of danger to Christianity should thus early be found vindicating some of the most peculiar and long contested doctrines of revelation. And yet it ought not to surprise us, for geology is as really the work of God as revelation. And though, when ill understood and perverted, she may have seemed recreant to her celestial origin, yet the more fully her proportions are developed, and her features brought into daylight, the more clearly do we recognize her alliance to every thing pure and noble in the universe."*

* President Hitchcock's Religion of Geology, pa 368.

372. Geology manifests the uniformity of nature's laws, carrying back their operation indefinitely into the past, exhibiting that unity of design which characterizes the present, extending through all periods of the world's history. It re-recognises the agency of those subtle powers, heat, light, electricity and chemical attraction, regulating all the changes that occurred in the constitution of bodies in all ages. The structure of all the fossil forms of animal and vegetable bodies, shows the prevalence of the present anatomical and physiological laws, in organic systems long since extinct, and links them all into one grand, harmonious system, worthy of the great Contriver. All the proofs of power, wisdom, and benevolence evinced at present, in the adaptation of animate beings to the circumstances in which they are placed by the Creator, are discernible in all periods of fossil botany and zoology. Their deviations from the present races in form or size do not render them anomalous or monstrous. "The animals of the antediluvian world," says Sir Charles Bell, the distinguished anatomist, "were not monsters; there is no *lusus* or extravagance. Hideous as they appear to us, and like the phantoms of a dream, they were adapted to the condition of the earth when they existed." This uniformity in the structure and correlation of parts of animated frames, and the adoption of analogous means for various ends, with such deviations only as the diversity of circumstances in which they were placed required, show the immutable wisdom and benevolence, as well the unity of design of the Creator.

373. Although geology does not account for the origin of matter nor aid us in forming a conception of its creation, it does exhibit modifications of it, whose production and regulation require the intervention of a Deity. The appearance

of the various vegetables and animals in successive periods can not be accounted for independently of creative power, and we know of no such power other than the Deity, since the development hypothesis is geologically demonstrated (§ 138) to be false.

374. The instances of special adaptation of means to ends with reference to the welfare of the present races of animate beings, particularly of man, are numerous. Such are the inequalities of the earth's surface, produced by energetic forces acting from below, which cause the circulation of water and prevent universal stagnation and death; the production of soils adapted to sustain vegetable life, by the violent agency which has disintegrated the rocks and commingled their fragments; the protrusion of metals from deep recesses to accessible positions in veins; and the accumulation and wide diffusion of the useful minerals, rock-salt, coal, marble, &c.

375. Geology coincides with other sciences in expanding our views of the grandeur of the Universe and the plans of the Deity. The microscope discloses to us myriads of living beings in a drop of water, the number increasing with the power of the instrument. The telescope reveals the existence of innumerable suns at such distances from us that their light—though traveling at the rate of two hundred thousand miles per second—requires thousands of years to come to our world. So Geology, instead of limiting our contemplations of the history of our globe to the past six thousand years, carries us back into the indefinite past, "developing a plan of the Deity respecting its preparation and use, grand in its outlines, and beautiful in its execution."*

376. The correct interpretation of the Mosaic account of

* Hitchcock.

the Creation has been the subject of much discussion. Until recently, the commonly received opinion respecting it has been, that it taught that the universe began to exist about six thousand years ago, and that its creation was accomplished in six literal days. Geology teaches that the world has existed for an indefinite period, much longer: hence arises an apparent discrepancy, and it becomes eminently desirable to ascertain which interpretation is correct. Much of the difficulty on the subject has arisen from the peculiarities of style and modes of description used in these ancient writings, which are not only not such as are used in scientific treatises, but not such as accord with the state of knowledge and prevalent opinion of the present day. The writers of the Old Testament, says Dr. J. Pye Smith, use "language borrowed from the bodily and mental constitution of man, and from those opinions concerning the works of God in the natural world, which were generally received by the people to whom the blessings of revelation were granted." They describe natural objects and events as they appear to the eye, which is not in accordance with their real nature; they speak, says Rosenmuller, the eminent German commentator, "according to optical, and not physical truth." Similar discrepancies occur in the scriptures with reference to astronomical, physiological, and chemical phenomena. The Creator did not design to anticipate the discoveries of science, and correct the erroneous views entertained on these subjects by the people to whom the revelation was addressed.

377. Several attempts have been made to correct the interpretation of the Mosaic account of the creation, so as to make it accord with the established rules of philology and the facts proved by geology. Nor are attempts of this kind confined to this topic. Commentators on ancient writings

seek all the light which science, history and antiquities shed upon the subject of their investigations; in repeated instances have modern discoveries, in geography, botany, mineralogy and other branches of science, essentially modified the interpretation of passages in such writings. The same term may convey opposite meanings to different readers. The terms *elements* and *combustion* had for the ancient Jew an import different from that which the chemist derives from them. A term is used in the twenty-second verse of the second chapter of Jeremiah and in the twentieth verse of the twenty-fifth chapter of Proverbs, which is translated in the English version *nitre;* the nitre of modern chemistry is the nitrate of potassa, which would render the passages cited unmeaning: but if the nitre of the Jews was the carbonate of soda, the term is apposite and forcible. The propriety of such use of science in interpretation is sanctioned by philologists. "If I am reminded, in a tone of animadversion, that I am making science in this instance the interpreter of Scripture, my reply is that I am simply making the works of God illustrate his Word in a department in which they speak with a distinct and authoritative voice; that it is all the same, whether our geological or theological investigations have been prior, if we have not forced the one into accordance with the other."* Dr. Harris also, in his Pre-Adamite Earth, remarks, "it might be deserving consideration, whether or not the conduct of those is not open to just animadversion, who first undertake to pronounce on the meaning of a passage of Scripture, irrespective of all the appropriate evidence, and who then, when that evidence is explored and produced, insist on their *a priori* interpretation as the only true one."

* Davidson's Sacred Hermeneutics quoted by President Hitchcock.

378. Assuming, as to every geologist seems reasonable, that the period of about sixteen hundred years intervening between the Creation and the Flood was inadequate to the production of the stratified rocks including the remains of extinct races of plants and animals; and that the deluge was too short and entirely unfitted in its nature, to produce those rocks, we are reduced to two principal modes of reconciling the apparent discrepancy between the Mosaic and Geological accounts.

The first of these interprets the word *day* in Genesis as a period of indefinite time, during which several geological formations may have been perfected. The advocates of this view urge that such a sense is attached to the word in many languages in use, and that it was so used repeatedly, as is shown by contexts, in the Old and New Testaments, even in the chapter of Genesis, applied to the subject in question: "These are the generations of the heavens and of the earth when they were created in the day that the Lord God made the earth and the heavens, and every plant of the fields." They also argue that the order of creative acts revealed in the sacred record harmonizes with that developed by geological researches.

It is objected to this interpretation that most of the fossil races had become extinct when those at present existing were introduced, so that if Moses described the fossil races, those that now exist must have been created with man on the sixth day, which does not accord with the sacred records, and no reason is shown why the remains of the existing races, if they were concomitant with the fossil, were not preserved with them. It is also objected that while there is a general resemblance in the succession of events detailed in the two accounts, the coincidence in the order of introduction of the various animate forms is by no means accurate, since instead of finding the lower half of the fossiliferous strata containing vegetables only, we meet with a preponderance of animals.

379. The mode of reconciling the apparent discrepancy which is most generally received, regards the first verse ot Genesis as having no immediate connection with the following verses, but simply asserts that in the indefinitely remote past,—"in the beginning," God created the universe; then passing by an indefinite interval during which all the fossiliferous strata, up to those of the present period, were deposited, the subsequent verses give the account of the introduction of the present races; and that this renovation or remodelling, not creating, of pre-existing materials was in six days.

Some of the ablest expositors assert that this interpretation is in strict accordance with the rules of exegesis, and if admitted, the apparent discrepancy disappears.

Dr. J. Pye Smith has proposed to extend this interpretation, on the acknowledged principle that the sacred writers adapted their language to the limited knowledge of the Jews, and consequently did not imply by the term *earth*, the entire globe, but that portion of it known to the Jews, "which God was adapting for the dwelling place of man and the animals connected with him." This view would obviate to some extent a difficulty which arises from the fact that many of the races introduced in the tertiary period long before the creation of man still survive: some have supposed that these were destroyed and again created, which seems quite improbable. This hypothesis of Dr. Smith's coincides also with what Natural History teaches respecting the distribution of plants and animals, viz: that they have not dispersed from one centre, but have spread from many centres of creation.

380. If it be admitted that much remains to be accomplished before a perfectly satisfactory comparison of geological results with the sacred text is attained, still in the present aspect of the case it is unphilosophical to presume there is any real collision between them. "It is not necessary" says President Hitchcock "that we should be perfectly sure that the method which has been described, or any other, of bringing geology into harmony with the Bible, is infallibly true. It is only necessary that it should be sustained by probable evidence; that it should fairly meet the geological difficulty on the one hand, and do no violence to the language or spirit of the Bible on the other. This is sufficient, surely, to satisfy every philosophical mind that there is no collision between geology and revelation. But should it appear hereafter, either from the discoveries of the geologist or the philologist, that our views must be somewhat modified, it would not show that the previous views had been insufficient to harmonize the two subjects; but only that here, as in every other department of human knowledge, perfection is not attained, except by long continued efforts."*

381. It has been supposed that a change occurred in the constitution of men and animals at the time, and in consequence, of the apostacy by which they were rendered mortal. Such an opinion has been based upon the text—*By one man sin entered the world, and death by sin.* Geology, however, shows that death had occurred in innumerable instances before the creation of man, while physiology demonstrates

* This whole subject is most fully and satisfactorily discussed in President Hitchcock's Religion of Geology and its connected sciences.

that mortality is a universal law of organic beings, requiring miraculous interference to prevent its fulfillment. A comparison of Scripture texts clearly shows that the death referred to in them is limited to man:—*And so death passed upon all men, for that all have sinned.* This limitation to moral agents is also inferred from the text—*Since by man came death, by man came also the resurrection of the dead*

382. The belief in a deluge is very general among civilized nations, and such an event is explicitly described in the Bible. To the flood was formerly assigned the origin of all the stratified rocks, and still more recently the phenomena of the drift were ascribed to its agency. But Geologists are agreed that no phenomena can at present be identified as the result of the historical deluge, since the nature of the agency exerted differs, in several respects, and the period of the drift was antecedent to the creation of man. Since, however, every portion of the strata has been submerged, and usually many times, geology renders the occurrence of Noah's deluge eminently probable. Considerations derived from Natural History, the great number of species—probably five hundred thousand—and the amount of food requisite to sustain them during the prevalence of the flood, have led most writers who have investigated the subject, to the conclusion that the deluge was partial, not covering the whole earth, and it is admitted that the design of the penal infliction would have been subserved by such a flood, and that the Scripture writers are accustomed to use universal terms to signify large quantities.

CHAPTER XIV.

GEOGRAPHICAL GEOLOGY.

383. By Geographical Geology is understood the description of the structure of particular districts included within natural or political divisions. The geology of many countries is very imperfectly known, but important deductions respecting it may be obtained from their physical geography, since the connection between the geographical features of a country and its geology is very intimate. The forms and positions of *Mountain Chains* often furnish a satisfactory means of judging of the probable range and extent of certain formations, which have been carefully investigated in other localities. The geology of a region is usually complicate in proportion as its contour is broken and irregular, indicating that it has been subjected to numerous upheavals.

ASIA.

384. *Asia* and Europe form one continent, to which Africa may be considered as a peninsular appendage. The form of this great continent has been determined by the immense zone of mountains and table lands which, commencing on the coasts of Portugal and Barbary, on the Atlantic ocean, stretch through a distance of ten thousand miles, to the Pacific ocean, in China and Japan. Adjoining this belt on the north lies a vast plain, extending from the

Pyrenees to the remotest point of Asia, with few interruptions by transverse spurs: a similar plain on the south is indented by gulfs and arms of the sea. The desert called the Great Gobi is a plateau of six hundred thousand square miles, elevated more than four hundred feet above the sea. The axis of this mountain chain is granite, though the crest of the Himalayas is gneiss and other metamorphic rocks. The Silurian strata lie at an elevation of sixteen thousand feet above the sea, and much more modern strata are also found in elevated positions, showing that great geological changes have occurred over very extensive tracts of this continent, in comparatively recent geological periods. On a branch of the Altai range, between the rivers Obi and Yenissei, are extensive coal beds which, it is affirmed, were set on fire by lightning, and have continued to burn more than a century.* Some of these mountains surpass the Andes in the amount of their metallic products, especially of gold, which is wrought in the metamorphic slates near dikes of igneous rocks, and in the alluvium of these rocks.

385. In *Siberia*, the Silurian, Devonian, and Permian strata have been recognised. More than two hundred thousand square miles are covered by the gold alluvium; and in addition to gold, silver, platina, copper, iron, and a great variety of gems are found. The surface is generally undulating, but much of it is flat and low: interesting organic remains, as the mammoth and extinct rhinoceros, have been obtained from the frozen gravel.

Tartary is intersected by four systems of mountains with all classes of strata, from primary to tertiary, salt lakes, deserts, and very numerous volcanoes, extinct and active.

386 In *Thibet*, *Hindostan*, and *India*, the summits of

* Somerville's Physical Geography—page 60.

the mountains are of gneiss and mica slate, traversed by granite and porphyry; these are succeeded by talcose slate, limestones, coal, sandstones of the newer secondary series, and the tertiary strata, including numerous organic remains, among which are those of the mastodon, hippopotamus, deer, gavial, crocodile, sivatherium, and monkey. The unstratified rocks also abound in Hindostan, which is celebrated for its beautiful gems, especially the diamonds of Golconda, of surpassing brilliancy and hardness, which are found in the conglomerate and alluvium. *Ceylon* consists chiefly of stratified primary rocks, and is celebrated for its gems.

South-eastern Asia consists of a range of mountains of plutonic rocks, partially covered by slates, sandstones, and alluvium; they contain gold, silver and arsenic, and constitute the richest repository of tin known on the globe. In *Sumatra*, *Java*, and *Borneo*, volcanic mountains occur, and at their bases secondary and tertiary strata. Their mineral products are diamonds, gold, tin, copper, and sulphur.

387. The geology of *China* is varied, embracing all classes of rocks, abounding with metals and precious stones. Its alluvial formations are very extensive. In *Japan*, coal, amber and sulphur are found, together with metals. Numerous volcanoes also exist there

388. Some portions of *Persia* are mountainous, while others are level; others still constitute a sandy desert. Unstratified rocks, palæozoic, secondary and drift deposits, have been identified. The white alabaster of Tabreez is a calcareous deposit made by thermal waters near Lake Oroomiah; the waters of the lake contain one-fifth of their weight of salts. Extinct craters, boiling springs, and deposits of sulphur and asphaltum indicate recent extensive volcanic agency.

389. The geology of *Arabia* also is varied by mountains, deserts and fertile plains. The rocks near the Red Sea and Arabian Gulf are granitic. Mount Sinai is syenitic granite. The valley of the Jordan is a fissure through which volcanic agency has been exhibited, and in which there is extensive subsidence below the level of the ocean. Mount Lebanon, which abounds with fossil fishes, belongs to the cretaceous formation.

390. *Polynesia* consists of a group or band of islands, volcanic, and coral reefs, which extend from the eastern coast of Asia, with some interruptions, to the west coast of America, intimating the existence of a prolongation of the great continent, but slightly depressed beneath the ocean level.

The rocky coasts of *New Holland* present granite, slates, limestones, sandstones, coal, salt, and Oolite. The fossils of its limestone caverns and osseous breccias belonged to animals closely allied to its present fauna.

EUROPE.

391. *European* geology is more thoroughly investigated than that of any other quarter of the globe, and many designations in the science have been derived from their local application in Europe.

The geology of the *British Islands* is more varied, and has been more accurately studied than that of any other area of equal extent in the world. It presents a nearly perfect succession of all the formations, and has been made the type or standard of reference for general geology.

392. The primary rocks of England are confined to its western portion, or Wales, where also are developed the Cambrian and Silurian formations. In passing from Wales

to London, the rocks rise in the series until we reach the center of the London basin, which is tertiary. The carboniferous system, embracing extensive deposits of coal, extends from the south of Wales to the Scottish border. The Newcastle beds have been most extensively wrought. The Welsh mines yield gold, copper, lead, tin, and iron; rock-salt and gypsum are found above the coal.

393. *Scotland* is more mountainous than England, and its rocks are principally primary and palæozoic: the islands near its coasts are of igneous origin, and present splendid basaltic columns and caves.

The rocks of *Ireland* are chiefly palæozoic, with some secondary formations: trap rocks cover an area of eight hundred square miles in the northern part of the island. Its mineral products are gold, copper, iron, coal and peat.

394. The geology of *France* presents nearly all the rocks stratified and unstratified. The secondary, especially the Oolitic limestone, and the tertiary series are conspicuous, but the carboniferous system is less developed than in England. The great platform of Auvergne was the theatre of violent volcanic action during the tertiary period, and many craters still exist, with perfect forms. The valley of the Rhine yields gold; while silver, copper, iron, and tin, are obtained in other localities; extensive quarries of gypsum, buhrstone, and flint are wrought. The surface of *Belgium* and *Holland* is very flat, and a portion of it is lower than the sea level. No unstratified rocks occur; the stratified rocks from the clay-slate are found, including some tertiary. Coal-beds and iron mines are wrought.

395. Primary mountains bound *Germany* on the east and south-west. All of the fossiliferous formations are represented, including the tertiary; the drift abounds, as

SECTION FROM THE NORTH OF SCOTLAND TO THE GULF OF VENICE.—(*Conybeare.*)

Fig. 167.

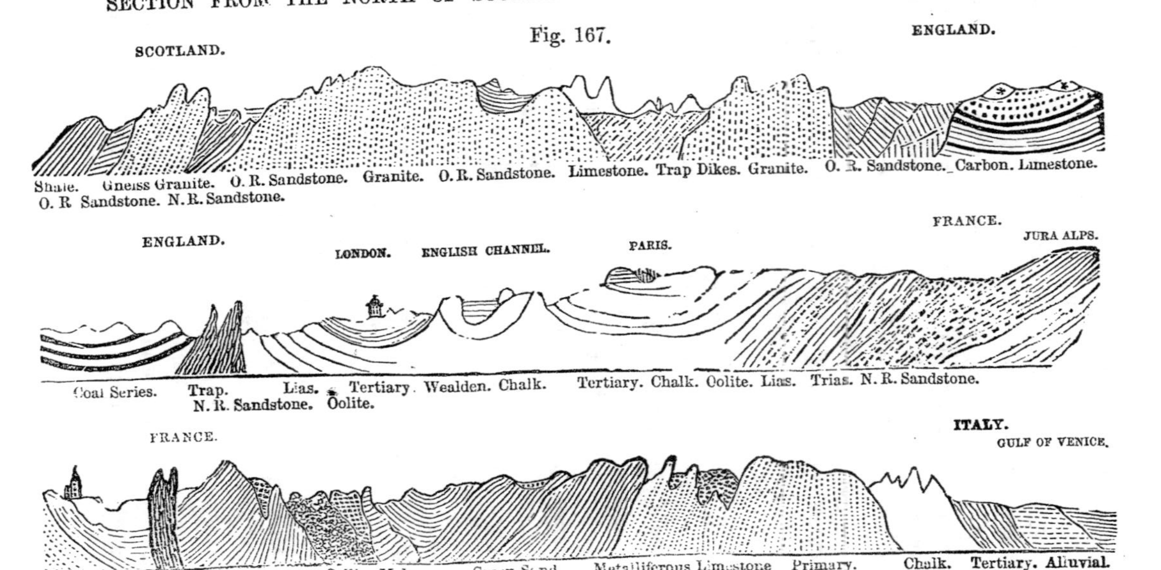

it does generally in Northern and Central Europe. The mineral products are numerous, presenting some of the best mining districts in Europe.

The geology of *Switzerland* is complicate: the central axes of the Alps, which are primary, are covered by the secondary and tertiary series, the latter in some instances at the height of four thousand feet above the ocean, showing that these mountains have been recently elevated.

396. In *Sweden* and *Norway*, the older rocks predominate: chalk and the tertiary are also found, but the drift presents the most striking features. Gold, silver, and copper are obtained, but the iron found in the gneiss is the most important metallic product. The more recent rocks, the wealden, chalk, tertiary, and alluvial, compose the surface of *Denmark*, while the igneous rocks, greenstone and lava, abound in *Iceland* and the *Faroe Islands.*

397. *Russia* and *Poland* are vast plains, bounded by mountains of primary rocks. The silurian, devonian, upper secondary and tertiary strata occur, covered to a great extent by the drift. Deposits of salt with gypsum are found in the permian and tertiary strata.

The numerous mountain ridges and vast plains of *Austria* present all the varieties of geological formations, and mineral products.

398. Northern *Italy* consists of extensive secondary and tertiary plains, sloping from the Alps; Mount Bolca is famous for its fossil fishes. The form of the peninsula depends upon the Apennines which are of limestone, while the sub-Apennine hills are of tertiary. The Apennines have been elevated several thousand feet since the tertiary period. The traces of volcanic agency are numerous in Italy. The marbles of Italy are celebrated for their beauty and variety.

The mountainous regions of *Spain* and *Portugal* consist of primary and secondary strata, the chalk being found at considerable elevation on the Pyrenees. Tertiary strata occur together with extinct volcanoes of that period. Rock salt is found in the cretaceous strata, and quicksilver in the clay slate.

AFRICA.

399. The Atlas chain of mountains separates the Mediterranean sea from the great desert; these mountains are composed of primary rocks, and the strata on their northern slope, in the *Barbary States*, are secondary and tertiary, into which trap rocks have been frequently intruded, and in which salt and gypsum are found.

Through upper *Egypt*, *Nubia*, and *Abyssinia*, occur primary rocks, granite, porphyry, syenite (so called from Syene) and limestone; drifting sands from the desert have encroached upon these territories, while extensive alluvial deposits have been made by the Nile in lower Egypt.

400. The western coast of Africa, for several degrees of latitude on either side of the equator, is composed of granite, syenite and the metamorphic rocks; the great quantity of gold formerly obtained here led to the designation of Gold coast.

Central Africa is traversed by the mountains of the Moon, which are primary and basaltic. Much of the gold obtained on the western coast was derived from the metamorphic rock of this range. The secondary series also, including the cretaceous formation, with rock salt, are found upon the Northern Slope.

401. The *Sahara* or great desert is a plain, slightly elevated above the ocean, extending from the rocky hills bounding the valley of the Nile to the Atlantic Ocean, two

SECTION OF AFRICA.—(FROM BOUE'S GEOLOGICAL MAP.)—Fig. 166.

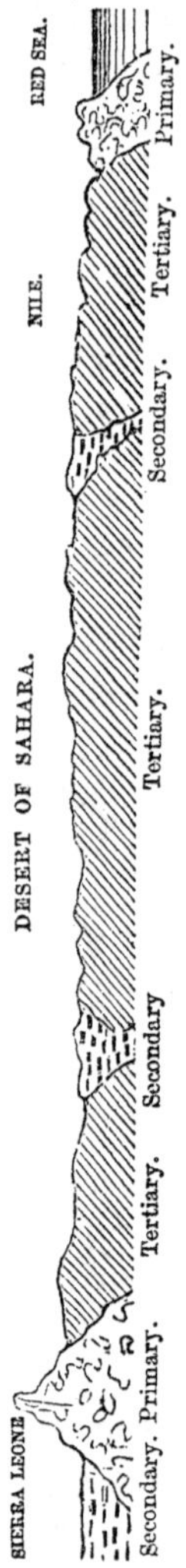

thousand six hundred and fifty miles in length and varying from seven hundred to twelve hundred miles in width. Its surface is loose sand, with intervening portions of gravel, pebbles, earth, and salt, and occasionally fertile spots—*oases*—watered by springs. No rain falls upon the desert. The chalk, tertiary rocks, salt, and shells identical with species still living in the Ocean, show that this is the bed of a great Mediterranean sea but recently elevated.

402. Chains of mountains consisting of primary and secondary rocks bound the two coasts of *Southern Africa,* between which intervenes an extensive plateau or table land, which merges at the southern extremity in the Cape Mountains; the tabular appearance of these Mountains at the Cape of Good Hope is due to horizontal masses of sandstone lying upon granite.

The islands in the vicinity of Africa, the Azores, Canaries, Cape Verde, St. Helena and Bourbon are of igneous origin; the volcanic peak of Teneriffe rises one thousand two hundred and seventy-five feet above the ocean. The axis of Madagascar is a chain of mountains, parallel to, and of the same age with the coast chain on the continent, the Mozambique channel only intervening.

SOUTH AMERICA.

403. The form of the *South American* continent is due to the position of the three mountain chains, the Andes running near the western coast from Cape Horn to the isthmus of Panama, a chain of small width but of majestic height, dipping rapidly towards the Pacific, but sloping on the east into level plains of great extent; the Brazil chain between the Rio de la Plata and the Amazon river; and the system of Parima and Guiana between the Amazon and Orinoco rivers. These mountains are primary and volcanic, covered by slates fossiliferous limestones and red sandstones of various geological ages. Coal and chalk are found at an elevation of thirteen thousand and fourteen thousand feet above the Ocean.

The extensive plains east of the Andes are so low, even near the foot of the Andes that a rise of one thousand feet in the Atlantic Ocean would submerge more than one half of the continent of South America. These plains are divided by the mountains and table lands of Parima and Brazil into three different basins differing in aspect: the *Llanos*, or grassy steppes of the Orinoco; the *Silvas* or woody basin of the Amazon covering an extent of two hundred and eighty thousand square miles; and the deserts and *pampas* of Buenos Ayres and

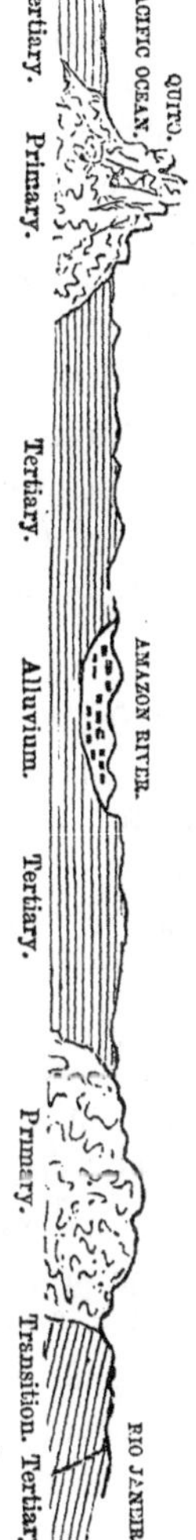

SECTION OF SOUTH AMERICA.—(*Boue.*)—Fig. 169.

Patagonia. These plains are of recent geological origin, and have furnished very interesting fossils. The tertiary strata are extensively developed in many localities, especially along the terraced coasts of Patagonia, and bordering the plains. Portions of the continent many hundred miles in length have been raised from beneath the ocean within the period of the shell-fish now living, which are found in the plains still retaining their colors. The volcanoes of this continent are among the most magnificent of the globe.

South America has long been distinguished for its mines of gold, silver and platinum. The diamonds of Brazil are obtained from the alluvium; other gems also are found in the rocks, as the topaz, emerald and sapphire.

CENTRAL AMERICA.

404. The Andes continue through *Guatimala* and *Mexico* in an irregular mixture of table lands and mountains, consisting of granite, gneiss and mica-slate, with a large admixture of lavas ancient and modern; secondary sandstone and limestone occur together with alluvial deposits along the coasts. Few regions of the globe rival this in intensity of volcanic action. The mines of Mexico which are in talcose and mica slates, transition limestones and porphyry, have yielded a large amount of gold and silver.

The mountains in the *West Indies* are similar to those of South America of which some of the ranges appear to be continuous; this fact together with the identity of the fossil remains of extinct quadrupeds, renders it probable that the West Indian Archipelago was once a part of the American Continent, the area of the Gulf of Mexico and the Caribbean sea having subsided at a recent geological period. Secondary and tertiary strata are also developed upon these islands attended by the drift and alluvium. The islands are still subject to violent volcanic action, especially earthquakes, and extinct craters are common. The Pitch Lake in Trinidad is three miles in circumference and of unknown depth.

NORTH AMERICA.

405. The general structure of North America is simple; its form is due to the position of two mountain chains—the Rocky mountains running north-west, and the Alleghanies northeast, including one of the most extensive basins in the world, embracing three millions two hundred and fifty thousand square miles. The Rocky mountains are composed of primary rocks, covered by sedimentary rocks of various geological ages, intersected by volcanic eruptions, though active volcanoes are principally confined to the northern part of the chain and near the Pacific ocean. At the base of these mountains on the east lies a sandy desert four hundred or five hundred miles wide. Rock salt and salt lakes are found in its vicinity. The coal formation with its fossils of tropical character is found at Melville Island in 74½° north latitude.

The auriferous deposits of *California* are alluvial, the detritus of sandstones, limestones and slates, especially talcose slate, intersected by quartz and porphyry. The gold is also found in veins in these rocks. Cinnabar the ore of mercury, silver, platinum, iron, tin and lead are known to exist in this locality.

UNITED STATES EAST OF THE ROCKY MOUNTAINS.

406. The Alleghany or Appalachian chain of mountains separates the great Mississippi Valley from the Atlantic slope; it consists of a series of from three to five parallel ridges with intervening valleys. Extending into *New Eng-*

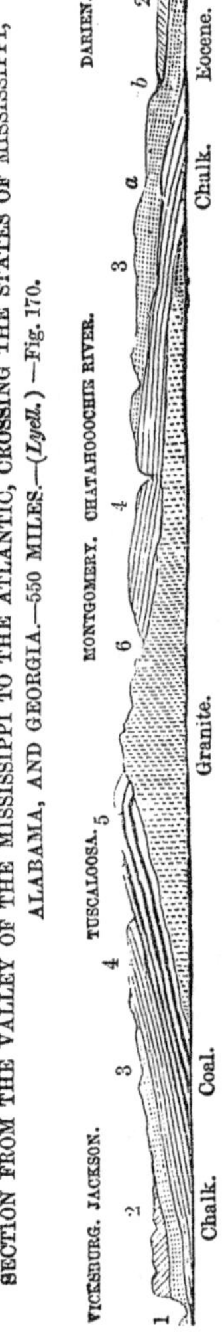

SECTION FROM THE VALLEY OF THE MISSISSIPPI TO THE ATLANTIC, CROSSING THE STATES OF MISSISSIPPI, ALABAMA, AND GEORGIA.—550 MILES.—(*Lyell.*)—Fig. 170.

land these mountains form the substrata of the geology of that region, of which the unstratified and metamorphic rocks form the principal part; limited portions of more recent sedimentary rocks overlie them, among which the new red sandstone is supposed to be recognized in the Connecticut Valley.

407. The Atlantic slope is very narrow at New York, but in its southern portion extends several hundred miles from the ocean. Upon this slope in New Jersey the new red sandstone is found, succeeded by the cretaceous formation, consisting of marls, limestones and green sands; the latter extend to Alabama. The coal in Virginia near Richmond is assigned to the Oolitic period. The tertiary series commencing on the coast of Massachusetts extend almost continuously along the Atlantic coast into the Mississippi Valley; in the Carolinas, Georgia, and Alabama, they are extensively developed, furnishing many characteristic fossils, which are however very rarely specifically identical with the tertiary fossils of Europe.

408. The tertiary rocks constitute the principal part of the surface in the Southern states; they repose upon the cretaceous, and in the lower portion of the Mississippi Valley, these together with

the alluvium overlie the palæozoic strata, as is shown in the section, Fig. 170 ; in which

1 indicates the modern alluvium of the Mississippi.

2. The ancient fluviatile deposit with recent shells and bones of extinct mammalia; loess.

2* Marine and fresh water deposits with recent sea shells and bones of extinct land animals.

3. The Eocene with remains of the Zeuglodon

a. b. Terraces.

4. Cretaceous formation, gravel, sand, and argillaceous limestone.

5. The Palæozoic—coal measures of Alabama.

6. Granite.

409. The stratified formations of the Mississippi valley and the western ridges of the Alleghany Mountains are the older—palæozoic rocks, which are expanded to a vast extent, and of very great thickness, while the secondary formations with the exception of very limited portions of the cretaceous, are deficient. Tertiary and alluvial deposits also extend up the valley of the Mississippi and its tributaries. The strata resting upon the primary rocks over large areas are the Silurian and Devonian. The carboniferous limestone is known to be widely extended in this valley, and the coal formation appears in many localities. These strata have frequently but little inclination, but are of great thickness, and abound in characteristic organic remains.

The identity of the great systems of the palæozoic rocks in Europe and in the Mississippi valley is easily recognised; that of the minor subdivisions is however, in many instances, obscure. The thorough geological surveys made in some portions of the United States whose rocks belong to these systems have led to the adoption of some provisional terms based upon certain peculiarities of the rocks, or upon the localities where they have been investigated. The subdivisions of the systems made by the New York survey, constitute, for the present, a convenient standard of reference. The following tabular arrangement, by Professor James Hall, exhibits the correspondence of these systems in Great Britain and New York.

TABLE.

SUBDIVISIONS OF THE ROCKS OF THE NEW-YORK SYSTEM.	SUBDIVISIONS OF THE SILURIAN AND OLD RED SYSTEMS IN GREAT BRITAIN.
Old Red sandstone	Old Red sandstone.
Chemung group. Portage group. Genesee slate. Tully limestone. Hamilton group. Marcellus shale.	Upper and Lower Ludlow rocks, including the Devonian System of Phillips.
Corniferous limestone. Onondaga limestone. Schoharie grit. Cauda-galli grit. Oriskany sandstone. Upper Pentamerus limestone. Encrinal limestone. Delthyris shaly limestone. Pentamerus limestone. Water-lime group. Onondaga salt group. Niagara group.	Wenlock rocks.
Clinton group. Medina sandstone. Oneida conglomerate. Grey sandstone. Hudson-river group.	Caradoc sandstone.
Utica slate.	Llandeilo flags.
Trenton limestone. Birdseye and Blak-river limestones. Chazy limestone. Calciferous sandrock. Potsdam sandstone.	These formations are not as fully recognized in Great Britain as in New York.

The order of succession in these rocks is exhibited in a section from Canada to Pennsylvania; from the primary to the Old Red Sandstone. Fig. 171.

410. The differences in character and amount of development which the same strata present in localities remote from each other, have induced a diversity of names applied to them in the United States. More extended surveys and comparisons will probably show the identity of some now regarded as diverse. Since the mineral characters of these strata are very fluctuating, their organic remains are deemed the most reliable means of determining their identity or diversity. A table similar to that just used for the comparison of British and American systems has been constructed by the same geologist to show the equivalency of the lower strata in America, bearing different names in five States, NewYork, Pennsylvania, Virginia, Ohio and Michigan.

Fig. 171

SECTION OF THE NEW YORK STRATA.—(*Hall.*)

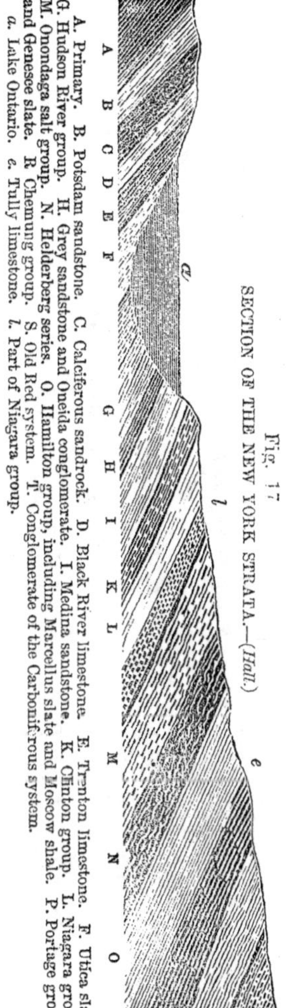

A. Primary. B. Potsdam sandstone. C. Calciferous sandrock. D. Black River limestone. E. Trenton limestone. F. Utica slate. G. Hudson River group. H. Grey sandstone and Oneida conglomerate. I. Medina sandstone. K. Clinton group. L. Niagara group. M. Onondaga salt group. N. Helderberg series. O. Hamilton group, including Marcellus slate and Moscow shale. P. Portage group. and Genesee slate. R. Chemung group. S. Old Red system. T. Conglomerate of the Carboniferous system.
a. Lake Ontario. *e.* Tully limestone. *l.* Part of Niagara group.

TABULAR ARRANGEMENT OF LOWER AMERICAN STRATA, SHOWING THE EQUIVALENCY OF THOSE KNOWN BY DIFFERENT NAMES IN THE SEVERAL STATES.

NEW YORK SURVEY.	PENNSYLVANIA AND VIRGINIA SURVEYS.	OHIO SURVEY.	MICHIGAN SURVEY.
Potsdam sandstone.	No. 1.	Wanting.	
Calciferous sandrock.	No. 2.	Wanting.	
Black River and Birdseye limestone.	No. 2.		
Trenton limestone.	No. 2.		
Utica slate.	No. 3.	Blue limestone and marl formation.	
Hudson River group.	No. 3.		
Oneida conglomerate.	No. 4.	Wanting.	
Gray sandstone.	No. 4.	Wanting.	
Medina sandstone.	No. 5.	Wanting.	
Clinton group.	No. 5.		
Niagara group.	Part of No. 6.	Part of Cliff limestone.	Mackinac limestone; gray, sandy, and porous limestone.
Onondaga salt group.		This formation is but partially developed in the south-west part of the State, but more extensively in the northern part.	
Water limestone.			
Pentamerus limestone.			
Delthyris shaly limestone.	Included in No. 6, if existing.		In this series are included the little Traverse bay limestone, Black bituminous limestone, Blue limestone, in thick, regular layers. Thunder bay limestone?
Encrinal limestone.		Wanting, or not recognized in the Western States.	
Upper Pentamerus limestone.			
Oriskany sandstone.	No. 7.		
Cauda-galli grit.	No. 7.		
Schoharie grit.			
Onondaga limestone.			
Corniferous limestone.		Upper part of the Cliff limestone.	Corniferous limestone.
Marcellus shale.	No. 8.	Black slate.	Shales. Black aluminous shale.
Hamilton group.	No. 8	Wanting, or but partially developed.	
Tully limestone.		Wanting.	
Genesee slate.		Wanting?	
Portage group.	No. 8?		Soft, light colored sandstones. Argillaceous slates and flagstones of Lake Huron. Sandstones of Point Aux Barques.
Chemung group.	No. 9.	Waverly sandstone series	

Fig. 172.

SECTION FROM WISCONSIN RIVER TO LAKE HURON.

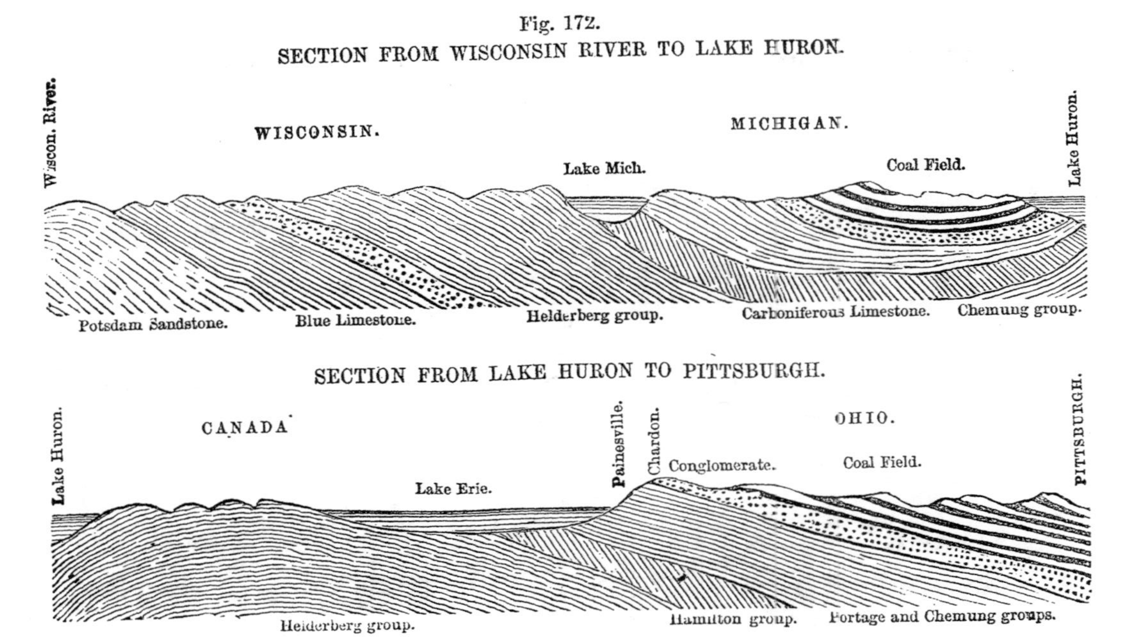

SECTION FROM LAKE HURON TO PITTSBURGH.

Fig. 173.

SECTION FROM ST. LOUIS THROUGH CINCINNATI AND WASHINGTON TO THE ATLANTIC OCEAN.

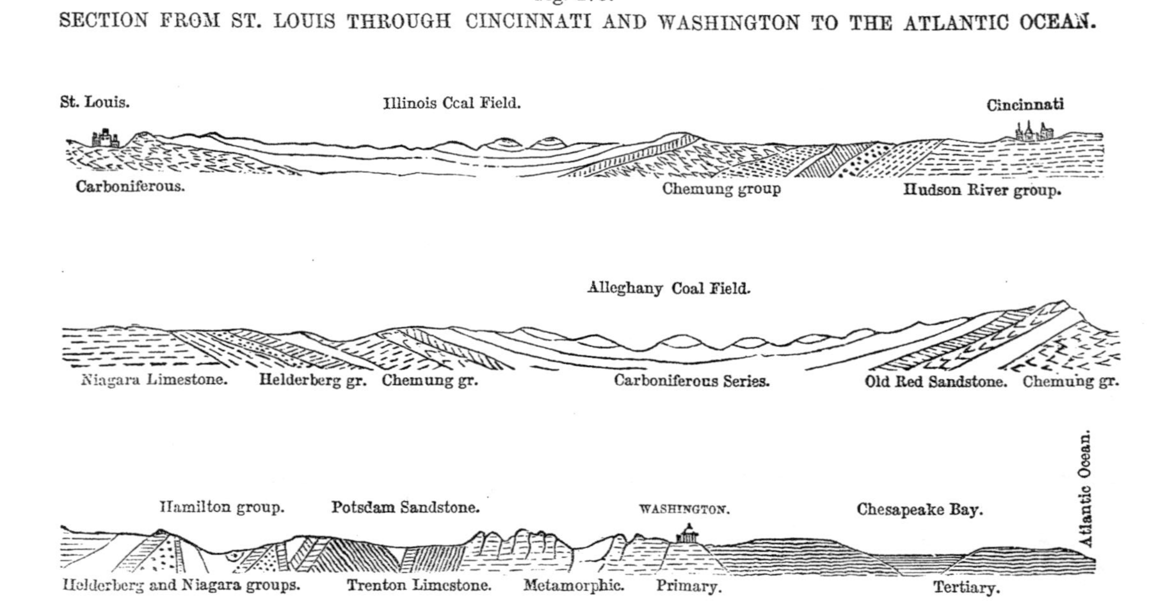

Fig. 174.

SECTION FROM THE PACIFIC OCEAN TO ST. LOUIS, MISSOURI.—(*Hall.*)

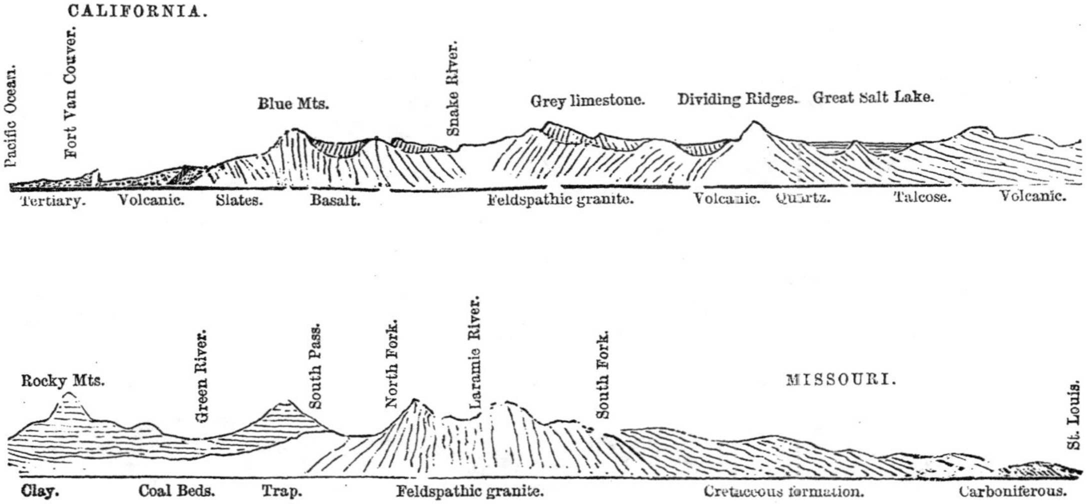

Fig. 175.

SECTION FROM PENNSYLVANIA TO NOVA SCOTIA.—(*Jackson.*)

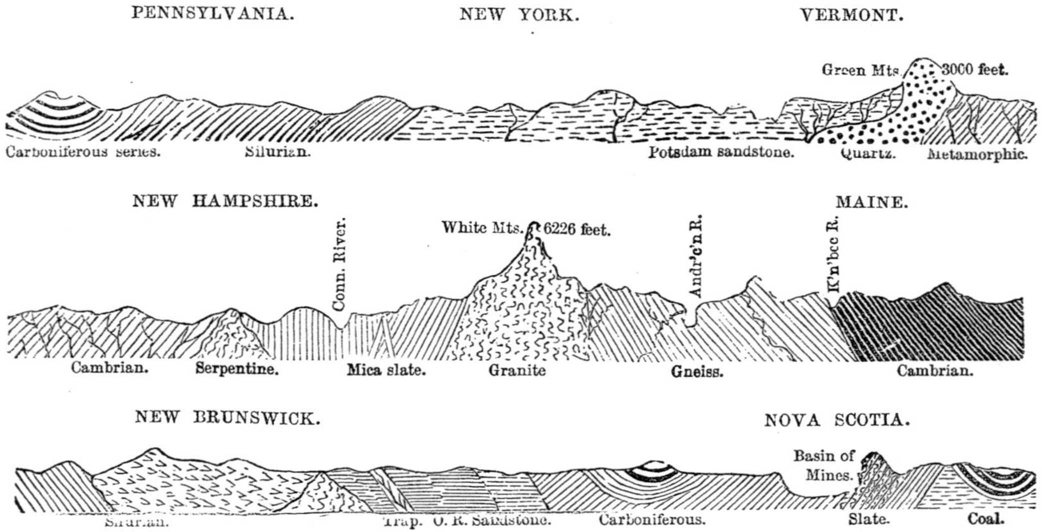

411. The *fossils* are said by Professor Hall to be more numerous in the rocks of America than in those of Europe of the same age. Some of the extinct species of the higher mammalia, found in the recent formations of special interest are, the *Mastodon*, *Mammoth*, *Megatherium*, *Megalonyx*, fossil *Elk*, several species of fossil *Ox*, *Walrus*, *Zeuglodon*.

412. The mineral treasures of the United States are, though imperfectly explored, known to be rich and abundant. The *Coal* measures of the Carboniferous period are developed on a scale, so far as is known, unequalled in the world, furnishing a supply of mineral fuel amply sufficient for the requirements of the whole civilized world, for thousands of years, even should the demand increase rapidly, and the consumption continue to bear reference to the multiplication of all kinds of industrial occupation.* The Alleghany coal field is seven hundred and fifty miles long with an average breadth of eighty-five miles, embracing an area of sixty-five thousand square miles, or more than forty millions of acres, distributed in eight states as follows:

Pennsylvania, - -	9,500,000	Kentucky, - - - -	5,750,000
Ohio, - - - - -	7,500,000	Tennessee, - - -	2,750,000
Virginia, - - - -	13,500,000	Georgia, - - - -	100,000
Maryland, - - - -	350,000	Alabama, - - -	2,250,000

The Illinois coal field also extends over an area of more than fifty thousand square miles; of which thirty thousand are in the state of Illinois, and eight thousand in Indiana. This is exclusive of the great coal-area of Missouri whose extent is not ascertained. The varieties of coal in these fields are bituminous, cannel, and anthracite; the latter is

* Ansted.

found in Pennsylvania occupying an area of two hundred thousand acres, in three principal districts; in one of which there are sixteen workable beds of three feet or more—the thickest being nearly thirty feet. The beds cease to be anthracitic at a distance from the primary rocks of the mountains.

The Michigan coal basin lies in the central portions of the State, as is represented by the shaded part of Fig. 176, and is one hundred and seventy miles long by one hundred miles wide. It is underlaid by carboniferous limestone, soft, light colored sandstones, argillaceous slate, and flagstones.

Fig. 176.

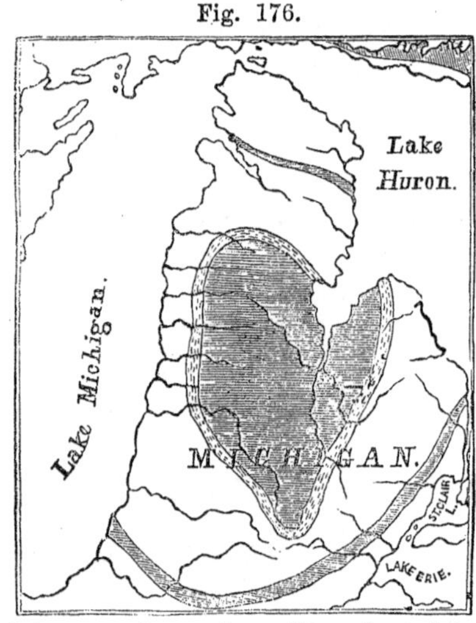

413. *Iron* is found in quantity and quality adapted to working in numerous localities. The deposits of specular iron in Missouri, and near Lake Superior are among the very largest in the world.

Copper occurs native and as an ore in great quantities on the shores of Lake Superior, in Missouri, and in less amount in Virginia, New Jersey, and New England.

One of the largest deposits of *lead* known in the world is wrought in the carboniferous limestone of the north-western and western states; and is estimated to be capable of yielding more than one hundred and fifty millions of pounds annually.

Gold is found in the primary rocks of the Alleghany Mountains in quartz rock traversing the metamorphic rocks, particularly talcose slate; and in greater quantities in the same geological position, and in the alluvium of these rocks, in California. *Silver* occurs also associated with gold, copper and lead. *Rock Salt* is found in limited quantity in the strata but is obtained in abundance from the brine springs, which rise from beneath the coal.

Inexhaustible supplies of granite, marble, freestone, and other rocks adapted to architectural purposes, are found in the primary rocks, but are as yet little wrought.

Mineral fertilizers of soils, *gypsum*, *marls*, and the *phosphate of lime*, are found in numerous localities. No active volcanoes exist in the United States, nor are there indications of igneous eruptions east of the Rocky Mountains since the protrusion of the trap. In the Rocky Mountains pumice and other evidences of more recent volcanic agency occur. Severe earthquakes are very rare; Thermal Springs however are abundant, and in some instances of a high temperature, as the hot springs of Arkansas and Virginia

GLOSSARY OF TERMS USED IN GEOLOGY.

NOTE.—Such terms as have been defined in the text are not repeated here, but may be referred to by means of the index.

AEROLITES. Mineral masses that fall from the atmosphere.

Albite. A white variety of Feldspar containing soda instead of potash.

Algæ. A division of cryptogamous plants, Sea weeds.

Alveola. Socket of a tooth.

Alveolus. A chamber of the belemnite.

Amorphous. Devoid of regular form.

Amorphozoa. Animals without definite form—Sponges.

Analcime. A simple mineral of the Zeolite family, found in trap and granite.

Analogue. A body resembling or corresponding with another body.

Anchylosis. A stiff, immovable joint.

Anhydrous. Without water.

Annelides. Worms having the integument formed of rings.

Antediluvian. Preceding the deluge.

Antennæ. Articulated horns of insects and crustacea—feelers.

Anthracotherium. An extinct quadruped, allied to the palœotheria found fossil in the coal beds of the tertiary.

Anthropomorphous. Resembling the human form.

Antiseptic. Preventing putrefaction.

Argentiferous. Containing Silver.

Arragonite. A variety of Carbonate of lime.

Asbestus. A fibrous mineral, of which an incombustible cloth is sometimes made.

Ashler. A name given to free stone when squared for building purposes.

Assay. The process of determining the amount of metal in an ore

Atom. The ultimate particle of an element.

Auriferous. Containing Gold.

Avalanche. A mass of ice, snow, or earth falling into a valley.

Asterolepis. A large fossil fish found in the Old Red Sandstone.

BACULITE. A many chambered shell resembling the Ammonite unwound.

Barytes. A mineral so called from its great weight.

Bassett. Outcrop of strata.

Bind. A miner's term for argillaceous Slate.

Blende. The Sulphuret of Zinc.

Bluff. A precipitous bank.

Botryoidal. Resembling a bunch of grapes.

Brachiopoda. A group of shell-bearing animals having two long spiral arms which assist in locomotion and in procuring food.

Byssus. A tuft of hairs by which some shell, fishes are attached to rocks.

CALAMINE. Zinc ore—the carbonate of Zinc.

Calc Sinter Calcareous deposits of Springs.

Calcine. To reduce to powder by heat; to expel carbonic acid as in burning lime.

Calp. An impure limestone of the palœozoic rocks.

Cannel-Coal. A hard bituminous coal burning with a clear flame and sometimes used instead of candles.

Carapace. The upper shell of some reptiles.

Carse. A Scottish term applied to the flat lands in valleys.

Cataclysm. A deluge.

Cetacea. Whales; vertebrated mammalia living in the water, but not fishes.

Chalybeate. Water holding iron in solution.

Chert. A silicious mineral resembling flint, but of coarser texture.

Choke-damp. Carbonic acid in mines and wells.

Clunch. The hard beds of the lower chalk.

Coleoptera. Insects having hard wing cases; beetles.

Conchoidal. Resembling a shell.

Conglomerate. A rock made up of rounded, water worn fragments cemented together.

Coombe. A dry valley.

Culm. An impure kind of coal.

Cumbrian. A name applied by Professor Sedgwick to a system of of rocks observed in Cumberland, England, now merged in the Cambrian or Silurian.

Cupriferous. Copper bearing.

DEBACLE. A great rush of waters, breaking down obstacles and dispersing detritus.

Debris. Ruins or fragments; detached from rocks, and accumulating in masses.

Denudation. Removal, by water, of masses overlying rocks, leaving them bare.

Dessication. Drying up.

Detritus. Sand, gravel. clay, &c., worn off from rocks by water.

Dolerite. A variety of trap-rock.

ECHINUS. A marine radiate animal covered with spines called sea urchin and sea egg.

Ecpyrosis. Destruction by fire.

Elytra. Wing-sheaths of beetles.

Epoch. A point in time from which a period is reckoned; also used as synonymous with period.

Equivalent. A term applied to strata in different regions, whose origin was contemporaneous.

Erosion. Wearing away by water.

Exuviæ. What is cast off; a name applied also to organic remains

FALUNS. A French term applied to some tertiary strata, resembling the English crag.

Facette. A little face.

Ferruginous. Containing iron.

Fiord. A deep, narrow inlet.

Fire Clay. A clay containing little alkali, and consequently very difficult to fuse.

Fire damp. Carburetted hydrogen in mines; mixed with air is very explosive.

Fissile. Easily split.

Fluviatile. Belonging to a river.

Flux. A substance used to render minerals more fusible.

Fuller's Earth. A compact, friable variety of clay, used in cleansing cloth.

Fulgurites. Vitrified sand-tubes which are supposed to have been produced by the striking of lightning on the sand.

GARNET. A simple red mineral, frequently found in mica-slate.

Glacis. A gentle slope, less steep than a *talus.*

Grit. A coarse grained sandstone.

Gyrgonites. Seedvessels or fresh water plants found fossil, and formerly supposed to be microscopic shells.

HABITAT. The district within which a species of plants or animals is naturally confined.

Hade. The dip or inclination of a mineral vein.

Hamite. A hook-shaped shell of an extinct cephalapod.

Homalonotus. A smooth-backed trilobite.

Hyphersthene. A variety of the mineral pryoxene, allied to augite.

IMBRICATE. Laid over each other like scales.

Iridescent. Shining with rainbow colors.

Isomorphous. Having same crystalline form with different chemical constitution.

KILLAS. A Cornish name for coarse slate.

LACUSTRINE. Belonging to a lake.

Leucite. A simple white mineral found in lava.

Lapilli. Globular volcanic ashes.

Laterite. Rock found in India, cut in the form of bricks and used for the same purpose.

Lithological. Pertaining to stones.

Lithodomi. Shellfishes that perforate and inhabit rocks.

Lithophytes. Stone plants; sometimes applied to corals.

Lithographic Stone. A slaty limestone used for drawing and printing upon.

Littoral. Belonging to the shore.

Loess. A tertiary deposit on the banks of the Rhine.

MARSUPIAL. Animals which carry their young in a pouch as the Opossum and Kangaroo.

Mesotype. A mineral of the Zeolite family often found in traprocks

Molasse. A soft sandstone of the meiocene period found in the great valley of Switzerland.

Moraines. A Swiss term for the debris brought into the valley by glaciers.

Moya. A term applied in South America to mud poured out from volcanoes.

NEACOMIAN. A group of rocks in the cretaceous system.

Nodule. A rounded mass.

Nucleus. A kernel, or point about which matter accumulates.

OCHRE. A yellow powder of earth and oxide of iron.

Oaplized-Wood. Wood petrified with Silica.

Ornithorhynchus. A genus of quadruped animals having the mouth like that of birds.

Ophite. A rock similar to serpentine.

Olivine. A simple, olive colored mineral found in basalt and lava.

Oryctology. A term formerly applied to the science of organic remains.

Osteology. The science which treats of bones.

Outlier. A portion of a stratum detached from the principal mass.

PACHYDERMATA. An order of quadrupeds having thick skins, as the Elephant, Horse, Pig. &c.

Pegmatite. A granite in which the three component minerals form distinct masses cemented together.

Pelagic. Belonging to the *deep* sea.

Phonolite. Clinkstone.

Phryganea. A genus of insects, whose larves, called worms, have been found fossil.

Pipe Clay. A plastic clay used in making pipes.

Pisolite. A stone possessing a structure like an agglutination of peas.

Pit Coal. Ordinary coal; so called because obtained from pits.

Plateau. A plain considerably elevated above the sea.

Puddingstone. Coarse conglomerate.

Pœcilitic. Variegated; name given in England to a part of the new red sandstone series.

Pyrites. A compound of sulphur and a metal so called because the chemical changes produce spontaneous heat; it is often mistaken for gold.

Polythalamous. Many chambered.

Protozoic. Exhibiting first forms of life.

Pryrogenous. Igneous; applied to melted rocks.

Pryroxene. A species of mineral including several varieties, augite, hypersthene, &c. of silicates of lime, magnesia, and iron.

QUADRUMANA. Four handed animals, as the apes.

RAG. A stone of coarse texture.

Rake-vein. A group of vertical veins.

Rubble. A term applied by quarrymen to the fragmentary masses of stone surmounting the beds.

Ruminantia. A group of quadrupeds including those which chew the cud as the ox.

SACCHAROID. Having the texture of loaf sugar.

Saliferous. Salt bearing; the new red sandstone called the saliferous system because it contains salt.

Salses. Eruptions of mud from small orifices in volcanic districts.

Savannahs. Low plains in North America, generally covered with wood.

Scaglia. An Italian rock of the cretaceous period.

Scarped. Having a steep face.

Schist. A slate not admitting of such perfect splitting as do the slates generally.

Seam. A thin bed.

Selenite. Crystallized sulphate of lime.

Silvas. The wooded plains of South America.

Sinter. A rock precipitated from water.

Slickensides. The smooth face of a fault; also one of the Derbyshire lead ores.

Slide. Miners name for a small fault.

Spar. A crystallized mineral which gives a ready fracture.

Stanniferous. Tin bearing.

Stoss. A Swedish term applied to the side of a rock struck or worn by the drift.

Suturbrand. A variety of wood-coal or lignite.

TABLE-LAND. Level land elevated above the seas.

Testacea. Molluscous animals having a shelly covering.

Turbinated. Shells which have a spiral or screw-form structure.

Turrilite. An extinct genus of chambered shells, allied to the Ammonite having the siphuncle near the dorsal margin.

UNDERLIE. The inclination of a mineral vein.

WACKE. A German term which has been applied to a soft earthy variety of basalt; it has been used quite indefinitely.

Warp. The deposit from muddy waters artificially introduced into lowlands.

Watershed. The line between two river basins; it is not always a mountain chain.

Whinstone. An English provincial term applied to some trap rocks

ZEOLITE. A family of simple minerals including several varieties usually found in the traps or volcanic rocks; the name is derived from their boiling up when they are exposed to the heat of the blowpipe.

ALPHABETICAL INDEX.

www.ingramcontent.com/pod-product-compliance
Lightning Source LLC
LaVergne TN
LVHW010202110826
845151LV00002B/580

* 9 7 8 1 4 2 5 5 3 5 9 7 1 *